THE AI
GENERATION

THE AI GENERATION

SHAPING OUR GLOBAL FUTURE WITH THINKING MACHINES

OLAF GROTH
AND MARK NITZBERG
WITH DAN ZEHR

PEGASUS BOOKS
NEW YORK LONDON

THE AI GENERATION

Pegasus Books, Ltd.
148 West 37th Street, 13th Floor
New York, NY 10018

Originally published under the title *Solomon's Code*

First Pegasus Books paperback edition August 2021

Interior design by Maria Fernandez

Library of Congress Cataloging-in-Publication Data is available.

ISBN: 978-1-64313-353-9

10 9 8 7 6 5 4 3 2 1

Printed in the United States of America
Distributed by Simon & Schuster
www.pegasusbooks.com

To our children, Hannah & Fiona Groth-Reidy and Henry & Cecily Nitzberg:

may the era of thinking machines empower your humanity.

CONTENTS

FOREWORD

by Admiral James G. Stavridis

Since graduating from the US Naval Academy, I have spent forty years in public service, analyzing opportunities and threats and leading large-scale efforts that seize the former and mitigate the latter. I consider myself fortunate to have served with many formidable colleagues in US and allied administrations as we navigated through the Cold War, the rising threat of terrorism in its aftermath, and the accelerating changes across the geo-political and geo-economic landscape in the years since. Turbocharged now by digital technologies, this constant process of global transformation has both enabled and threatened peace and prosperity across borders. We have seen technology facilitate a number of key shifts throughout recent history—from a bipolar to a multipolar global order surfacing myriad new actors; from an analog to digital economy yielding unrivaled degrees of connectivity and anonymity; from technology for the elites to technology for the masses enabling entirely new ways of educating and participating; and from physical labor to knowledge workers leading to new distributions in the economics of working and earning.

We now are living amidst a fifth seismic shift, moving from purely linear computing systems to more cognitive and human-like technologies. The application of neural networks and machine learning techniques has given computer systems the ability to learn with minimal supervision, recognize complex patterns, and make recommendations and decisions on our behalf. Some of these decisions are small, subtle, and merely convenient to our everyday lives; others will have major impacts on people around the world. The application of both types is increasing rapidly and driven often in stealth fashion, by both commercial and political actors, most of whom hold honorable intentions and want to make this world a better place. To a great extent, we do get a better world. But even so, far-reaching ripple effects and unintended consequences make these advanced cognitive technologies both a panacea and a Pandora's box.

As a Strike Group Commander more than a decade ago, I oversaw the use of remotely piloted aircrafts to conduct strikes throughout Iraq, Afghanistan, and the Horn of Africa. These systems were highly effective and reduced the odds of collateral damage. A few years later, as a four-star Admiral and NATO Commander, we used the same technologies extensively in Libya. That 2011 campaign recorded the lowest number of collateral damage issues of any major air battle in history. But in every one of those operations we kept a man or woman "in the loop." It was obvious to me at the time that, sooner or later, we would have to grapple with the crucial issues that emerge if or when we decide to take a human out of the loop. The technical, ethical, and moral issues involved are essential, and the debate that is now emerging has been a long time in the making.

Similar crucial discussions are unfolding around a wide range of AI-powered technologies, and we all need to engage in these deliberations. As these thinking machines analyze knowledge and synthesize new insights and wisdom, they become powerful command-and-control instruments in the hands of those who understand them. Those who don't understand these cognitive systems will face a decisive disadvantage, and that imbalance of power illustrates the dark side of this rapidly expanding field. But we must remember that power also lies in the ability of burgeoning AI models to synthesize various data streams and develop workable solutions for wickedly complex problems. Our remarkable human ability to develop technologies that augment the limitations of the human brain and all our human foibles is the bright side of this new cognitive era.

Consider, for example, the complex problem humanity faces in conserving natural resources and preventing climate change. Think about ways we might provide more equitable health care in the United States, where we all too often dismiss the problem as a zero-sum contest between dollars and benefits, regularly overlooking more holistic ways we could lead healthier lives. And what about diversity and productivity at work—a balancing act that seems so difficult we devolve into endless stereotyping and alienation? The sheer complexity of all these major challenges can overwhelm us, and we too often default to reductionist thinking to find quick and easy answers.

In our lives and careers, all of us must deal with the types of difficult problems that stretch our brains to their limits or beyond. My career brought the same, whether in my role as a destroyer captain, the supreme commander at the helm of NATO forces in Europe, the dean of one of the world's premier international leadership academies, or in my everyday life as a father and a husband. But what drives success is our ability, as adaptable and malleable human beings, to convert these challenges into new horizons for personal and societal growth. Here, the ability of artificial intelligence and cognitive computing systems to help us make decisions could provide vast benefits.

Imagine a sophisticated AI-driven system that could diagnose where the world's next food crisis will occur and then recalibrate the global supply chain to solve the bottlenecks preemptively. It turns out we're already working on exactly that. Envision AI-powered mental-health interventions that could sharply reduce suicide rates among the mentally ill. There are smart people working on that, too. We are already creating new technologies that help illiterate people partake in the economy through computer vision technologies. Developers have created ways to teach children languages, math, and other subjects in ways that are more fun, more individualized to their needs, and more effective. Some researchers are creating innovative ways to help employers motivate their employees by making their work more meaningful and providing them a deeper sense of purpose. Others are beating cyber terrorists at their game, using cognitive computing systems to predict attack patterns and react far faster to breaches that do occur.

These developments inspire me, but as someone who spent decades protecting and leading people, I know that power and trust can be abused if guided by the wrong values. One must navigate through dark places before reaching the light. That's why we need to shape the design and application of these cognitive technologies so they serve all humankind, rather than just the powerful and wealthy. Ensuring those beneficial ends, harnessing their potential and circumventing their destructive power relies on our ability to once again muster the world's talent and create a coordinated global effort. I encourage you to engage and become part of that formative endeavor.

We have lots of work ahead of us if we hope to ensure that these systems know us—and vice versa—before we allow them to guide, speak, or act for us. They need to be able to reflect on their own reasoning and, equally important, be able to explain their decisions and actions in ways that humans can understand, especially when life-critical decisions are at stake. They should include mechanisms that check and correct for biases in the data they use for analysis and learning. And they shouldn't be designed or operated to suppress others, discriminate against minorities, or cheat and take advantage of those who are less digitally experienced.

Solomon's Code is the first book I've read that fully illustrates how innovation in thinking machines is taking place around the world, and the different ways power, trust and values are playing out across societies. Olaf Groth and Mark Nitzberg establish a trailhead for our thinking, sketching out the likely pathways in which cognitive systems will influence our lives over the next ten to fifteen years. The authors paint a vision of the grand possibilities and what we have to do to achieve them, but they don't shy away from the perils and pitfalls we'll have to navigate along the way. They illuminate a promising future full of interesting, challenging dilemmas, but they stay away from unrealistic and oversimplified utopian or dystopian visions.

That is the kind of responsible leadership that works with soldiers in the field of battle, with stakeholders in the economy, with students in educational settings, and between all of us as everyday citizens in civic life. The people who design and control the new generations of thinking machines will need to embrace the same kind of leadership if we want to make it to the next horizon of human growth.

—Admiral James G. Stavridis, author of *Sea Power:*
The History and Geopolitics of the World's Oceans,
former Supreme Commander at NATO and
Dean of the Fletcher School at Tufts University

Introduction

The once-grandiose tales of artificial intelligence have become quotidian stories. Before the robots started to look and sound human, they automated real jobs and transformed industries. Before AI put self-driving cars and trucks on the highways, it helped find alternate routes around traffic jams. Before AI gave us brain-enhancing implants, it gave us personal assistants that converse with us and respond to the sound of our voices. While previous plotlines for AI promised sudden and sweeping changes in our lives, today's AI bloom has delivered a transformation one step at a time, not through an apocalyptic blowout.

Artificial intelligence now pervades our lives, and it's not going away. Sure, the machines we call "intelligent" today might strike us as rote tomorrow, but the tremendous gains in computing power, availability of massive data sets, and a handful of engineering and scientific breakthroughs have lifted AI from its early Wright Brothers days to NASA heights. And as researchers fortify those underlying elements, more and more companies will integrate thinking machines into their products and services—and by extension, deeper into our daily lives. These and future developments will continue to reshape our existence in both radical and mundane ways, and their ongoing emergence will raise more and new questions not only about intelligence, but also about the very nature of our humanity. Those of us not involved in the field can sit passively by and let this unfolding plot carry us wherever it leads, or we can help write a story about beneficial human-machine coexistence. We can wait until it is time to march in the streets, asking governments to step in and protect us, or we can get in front of these developments and figure out how we want to relate to them. That is what this book intends to do: To help you, the reader, confront some of the societal, ethical, economic, and cultural quandaries that an increasingly powerful set of AI technologies will generate. The following chapters illustrate how AI will force us to consider what it means to be intelligent, human, and autonomous—and how our humanity makes us question how AI might become capable of ethical, compassionate decision-making and something more than just brutally efficient.

These issues will challenge our local and global conception of values, trust, and power, and we touch on those three themes throughout *Solomon's Code*. The title itself refers to the biblical King Solomon, an archetype of wealth and ethics-based wisdom but also a flawed leader. As we create the computer code that will power AI systems of the future, we do well to heed the cautionary tale of Solomon. In the end, the magnificent kingdom he built and ruled imploded—largely due to his own sins—and the subsequent breakup of his realm ushered in an era of violent unrest and social decline. The gift of wisdom was squandered, and society paid the price. Our book takes the position that humanity can prosper if we act with wisdom and foresight, and if we learn from our shortcomings as we design the next generation of AI. After all, we are already dealing with new tests that these advanced technologies have presented for our values, trust, and power. Governments, citizens, and companies around the world are debating personal-data protections and struggling to forge agreements on values of privacy and security. Stories about Google's Duplex placing humanlike automated calls to make reservations with restaurants or Amazon's Alexa accidentally listening in to conversations have people wondering just how much trust they can put in AI systems. The United

States, China, the European Union and others are already seeking to spread their influence and power through the use of these advanced technologies, accelerating into a global AI race that might help address climate change or, just as easily, lead to even more meddling in other countries' domestic affairs.

Meanwhile technologically, as these systems gain more and more *cognitive* power, they might begin to reflect a certain level of what we would call *consciousness*, or the ability to metareflect on their actions and their context. We all win if we can first instill a proper *conscience* in AI developers and the systems they create, so we ensure these technologies influence our lives in beneficial ways. And we can only accomplish this by joining forces, engaging in public discourse, creating relevant new policy, educating ourselves and our children, and developing and following a globally sourced open code of ethics. Whatever pathway the future of AI might take, we must create an inclusive and open loop that enables individuals and companies to increase productivity, heighten professional and personal satisfaction, and drive our progressive evolution.

Humanity's innate and undaunted desire to explore, develop, and advance will continue to spawn transformative new applications of artificial intelligence. That genie is out of the bottle, despite the unknown risks and rewards that might come of it. If we endeavor to build a machine that facilitates our higher development—rather than the other way around—we must maintain a focus on the subtle ways AI will transform values, trust, and power. And to do that, we must understand what AI can tell us about humanity itself, with all its rich global diversity, its critical challenges, and its remarkable potential.

1

Where Human Meets Machine

People move through life in blissful ignorance. In many ways, our bodies and lives work like a black box, and we consider it a surprise misfortune when disease or disaster strikes. For better or worse, we stumble through our existence and figure out our strengths and weaknesses through trial and error. But what happens as we start to fill in more and more of the unknowns with insights generated by smart algorithms? We might get remarkable new ways to enhance our strengths, mitigate our weaknesses, and defuse threats to our well-being. But we might also start to limit our choices, blocking off some enriching pathways in life because of a small chance they could lead to disastrous outcomes. If I want to make a risky choice, will I be the only one who has agency over that decision? Can my employer discriminate against me because I decided not to take the safest path? Have I sacrificed an improbable joy for fear of a possible misfortune?

And what happens to us fifteen years from now, when AI-powered technologies permeate so many more facets of everyday lives?

The chimes from Ava's home artificial intelligence system grew louder as she rolled over and covered her head with the pillow. Despite her better judgment, not to mention the constant reminders from her PAL, she'd ordered another vodka tonic at last call. She already hated this day—the anniversary of her mother's diagnosis thirty years earlier—but the hangover throbbing in her temples was making this morning downright painful. The blinds rising and the bedroom lights growing steadily brighter didn't help. "Yeah, yeah. I'm up," she growled as she steadied herself with a foot on the floor. Slowly, she rose and walked toward the bathroom, her assistant quietly reminding her of what she couldn't put out of her mind this morning no matter how hard she might try: precancer screening today.

So far, the doctors and their machines hadn't seen any need for action. But given her mother's medical history, Ava knew she carried an elevated risk of breast cancer. *I'm only twenty-nine years old*, she thought, *I shouldn't have to worry about this yet*. Her mother was pregnant with Ava when she got her diagnosis, so it came as a complete shock. Her parents agonized over what to do—about the cancer and the baby—until they found a doctor who made all the difference. In the two decades since, progress in artificial intelligence and biomedical breakthroughs had eliminated many of the worst health-care surprises, and it seemed like medical science conquered a couple new ones in the few years since. Ava was old enough to remember when AI could identify and predict ailments half the time. Now, it hit 90 percent thresholds, and most people trusted the machine to make critical decisions (even if they still wanted a doctor in the room).

Ava snapped back into focus: "Where are my goddamned keys?"

A patient, disembodied voice reminded her: "You left your keys and sunglasses on the kitchen counter. I'll order a car for you."

She winced. "I gotta change your speech setting," she said. "You still sound too much like Connor." No time now. She headed out the door for the doctor's office. If she could, she would skip the precautionary checkups, but then she'd lose her health insurance. So today, she just had to go through the motions and the machines.

Just don't say that word. A couple hours later, seated in the consultation room, the pounding in Ava's head finally faded. The anxiety didn't. "Sorry about the delay," her doctor said as she breezed in and sat down.

"Everything looks fine for now, but we're starting to see some patterns in your biomarkers. The WellScreen check says about 78 percent of patients with your combination of markers and genetic disposition develop cancer within a decade. It's about time we look at some preventative measures."

The rational part of Ava's mind expected the news; she always knew how likely it was given her family history. Still, that word—she stared blankly for a moment while the jolt of "cancer" started to ease. The doctor leaned forward and put a hand on Ava's forearm, and Ava remembered again why she kept coming back to her. "There's plenty of time between now and any potential tumors that might come of this," the doctor said. "We have a lot of options."

Ava started to breathe a little easier and tried to convince herself this was a good thing: All the tests and machines caught this long before she ran short on options. The doctor nodded to the screen on the wall, and up popped Ava's customized patient portal. They waited a few seconds for the health monitor on Ava's wrist to connect and upload her real-time biodata. There was her ex-boyfriend again, Connor's voice oddly reassuring this time: "Do you want me to grant the patient portal access to your data?"

The recommendations filled the screen, and the doctor started explaining. "Do you plan to have children? Research shows that pregnancy-related hormones can be a strong defense against the development of some breast cancers. If you have a child in the next ten years, the best models suggest your chances of developing cancers drop to about 13 percent. But given the types of cancers that run in your family, we also have a tailored set of hormone therapies you could choose from. Better, because you're starting now; it won't be nearly as harsh as the old hormone drugs your mom dealt with. There are some side effects, but they're pretty mild in most patients. On their own, they'll drop your chances of developing cancer to less than 20 percent. If you do either of those—the child or the hormone therapy—and replace your biological breasts with Lactadyne prosthetics in the next eight years, your chances of breast cancer are essentially nil."

The doctor noticed Ava's furrowed brow. "Look, you don't need to decide right now. Take a little time to think about it. You can go through all these options anytime on your portal. Meanwhile, take a couple days to look at this waiver, too. If you're up for it, we can start collecting data about your home and habits through your health manager and its connection to your refrigerator, toilet, and ActiSensor mattress—the usual stuff like environmental quality, diet, and exercise. It weirds some people out, but the monitoring can help suggest simple lifestyle changes to improve your odds. Once we get that data, we can adjust some parameters in your daily life and think about how we might change your diet and exercise. You're on Blue Star insurance, right? Their Survive 'n' Thrive incentive plan offers some great rewards for people who do environmental and health monitoring. You give up some control—but, hey, it is your health we're talking about, after all."

The doctor chuckled. Ava shuddered. She didn't much care for any of the options. The whole ride home she wondered how much the diagnosis would sidetrack her dreams. *Will my predisposition toward cancer disqualify me from the Antarctica trip? I'm comfortable, but I don't have piles of money—will the insurance company raise my premiums if I wait a few years before starting preventative therapies? What if my bosses find out? Will they ask me to leave or take a different role? And what about Emily? Should I tell my love, my partner that I might develop cancer unless we adjust our lifestyle? Will she still want to buy the condo on the hill?*

Will Emily leave? Ava harbored no illusions about the program the doctor described. Sure, it would keep her clock ticking, but it also meant giving the machine a lot of control over her life, and she would pay a financial and personal price if she didn't comply. As her PAL started describing the program during the ride home, it began to dawn on Ava just how comprehensive this program would be. The insurance company and the doctor would put together a total treatment plan that would consider everything from her emotional well-being to the friends she hung out with. Her girls' nights out would never fly, at least not to the heights it did last night. Would she have to rethink her entire social life, her friends, and her schedule to make healthier choices? She might have to change her home environment to maximize the hormonal therapy. She might have to reduce the stress of her job, maybe even change jobs altogether.

Her mind was racing now: *Will I have to give up Ayurveda because it's not scientifically proven to minimize my risk? I could move to Germany, where regulators accept Ayurveda and allow personal AIs to integrate data from Indian and Chinese medicine. Emily loves traveling, but we never thought about living overseas . . .*

"Too much," she whispered. "It's too much." She took a deep breath and massaged her temples. "I can't go home right now, and I sure as hell can't concentrate on work." Her fingers trembled as she rifled through her purse to find her PAL. She chuckled about the device in her hand—whenever she needed a human touch, she relied on a machine to deliver it.

"Ava! What's going on?"

"Mom?" she said, her voice cracking.

Ava's PAL had made the call automatically, a sensor in the earpiece picking up on her anxiety through the minute electromagnetic impulses in her brain and skin. The PAL instantly correlated the best person to call in her current state—always Mom or Dad, at least when Dad wasn't gallivanting through some far-off place—which, of course, her PAL also knew to be the case during this time of the year. Ava couldn't even recall whether she'd acknowledged a prompt to connect the call. Sometimes the PAL would just call automatically, as she had set it to do at especially stressful times.

Normally, the chipper greeting and the background noise that came over her mom's eight-year-old iPhone drove her nuts, like an old vinyl record. Today, it couldn't have been more comforting. "Papa sends his best," her mother said. "He's teaching today in Shanghai. He said he got you the gift of the century. I told him I don't even want to know."

"He got the beacon, too?"

"Of course, honey. You haven't delisted either of us yet. And you better not, either. You need your Swarm."

"Yeah, I suppose," Ava said, trailing off. Every week, Ava's PAL asked if she wanted to switch her alerts from the Swarm of friends and family to only Emily, who still couldn't understand why Ava wouldn't make the change. Ava tried to explain her relationship with her mom, the peace she got from the idea of multiple loved ones responding whenever she needed a comfort call or a reassuring holo-message. But Ava had substituted Connor for the Swarm back then, and the fact that she wouldn't make the change now really burned at Emily.

Mom's voice snapped Ava back again: "So, Zut! in Berkeley, then? At least that's what my phone says."

Connor's disembodied voice piped in: "Fifteen minutes until we arrive at Zut!"

Their favorite lunch spot. The restaurant they'd gone to for years. An easy drive from the city. The fact that Zut! just popped up as her new destination didn't even register with Ava, though she hadn't been there in months. Still, her PAL assigned it a unique rating, based on voice diagnostics and states of mind. Ava still marveled about how often Connor's voice suggested the perfect place, just like he used to.

When they met at lunch, Ava couldn't stop hugging her mom. There was nothing like the real thing, communing with another body and all its warmth, tenderness, and vulnerability. It didn't matter that her mom's advice was pretty much exactly what Ava's doctor and AI recommended. The emotional connection and the depth of familial love imbued it with so much more credibility. *The AI knew,* Ava thought, *but mom* knows.

"I survived," her mom said, "and so will you. Your chances are so much better now, and at least you can take a little time to map out a more predictable path. God, I'll never forget how shocked I felt when the first doctor told us to terminate the pregnancy."

Tears started to well in Ava's eyes, but her mom pushed on: "Honey, there was no way I was going to let that happen. No way. When we talked to the second doctor, he realized that was nonnegotiable and looked for alternatives. It helped that he worked at a Catholic hospital, but I think he just understood the emotional side of it, the fact that fighting for something I so desperately wanted, motherhood, probably helped my chances." Her mom shook her head, sighed, and wiped away a tear. Her eyes bored into Ava's. "There's nothing worse than someone or something telling you that you have no options—especially when they might be wrong. You need to take care of yourself, but you need to live your own life, too."

Ava looked up at the hills and smiled as she rode toward her film studio. Mom and Dad won't be around forever, at least not physically, she mused, but something about the song selection during the drive reminded her of how intimately her PAL picked up on the little recordings, notes, conversations, and subtle guidance her parents always provided. Melody Gardot's "Who Will Comfort Me" piped in, followed by Con Funk Shun's "Shake and Dance With Me." Dad's favorites were ancient, but her PAL correlated her mood with data on her interactions with him and found the exact combination of empathy and pick-me-up she needed. "Go get the day," the message on her PAL said. She didn't even bother to check if her dad actually sent it, or if her AI just knew to post his favorite exhortation. She smiled again, soaking in the energy of the sunny day.

At the studio, the walk to her desk always prompted a sense of gratitude. She had initially accepted a Wall Street job, opting for the money and the excitement without ever consulting the Career Navigator. Had she never bothered to take her mom's advice and consult her old AI assistant before moving to New York City, who knows how many miserable years she would've spent at that investment bank?

Fortunately, the Navigator homed in on her passion and predisposition for all things living and environmental, despite her best efforts to convince even herself otherwise. Career advisers, with their engrained biases and imperfect data, had told her she was a science ace, so the recommendation seemed to fit. It was definitely better than investment banking, anyway. She embarked on a mission to help mitigate climate change, enrolled in social justice programs, and spent a year as a park ranger in Tanzania. It was a fantastic time, but she never felt fully satisfied by the work. She spent a year debating herself until her AI finally projected a life picture that truly excited her. That beautiful, lifelike hologram of her work—not so different from the studio she stood in today—eventually took her to NYU's writing and directing program. The first night out with her classmates, the night she noticed Connor sitting by himself at the end of the bar, she actually kissed her new PAL.

Her job changed dramatically in the years since. AI generated increasingly precise insights about audience consumption patterns, societal mood swings, and political trends. Now the studio's AI capabilities distilled narratives that guided plot development and created meaning for people in their daily lives. Ava would guide those narratives and enrich them with emotional content, imaginative imagery, and storyboards that spoke to the human mind and spirit—however undefinable that still was in 2034.

But not everyone integrated well when the studio, like so many other companies, installed deeper AI systems. Ava had a number of friends who started and aborted careers in different fields—accounting, civil engineering, and pharmacy majors who suddenly discovered their education had not prepared them for the days when machines would conduct analyses, calculations, and highly routine or repetitive tasks. Ava recalled all too well the many long nights of whiskey-induced commiseration with struggling friends. Yet, it had been the same sort of AI insights that set her on the right path.

Ava started flipping through the storyboards for the studio's next animated feature, occasionally stopping to dictate a few ideas. Each time she felt especially inspired by a change, she'd reload the entire package and start reviewing the fully revised plotlines from the beginning. Today, though, she just couldn't connect with the stories. Leaning back in her chair, she flipped her PAL to attention mode. Peso, her financial advice AI, immediately beeped in: "Hi, Ava. Looks like the markets will rebound tomorrow. We're picking up on improving geopolitical, productivity, and climate forecasts for next quarter. I give it a 75 percent probability, and we still have some time to move. Shall I put $2,500 of your savings into the market? Your medical and communication data suggest you'll be cutting back on consumables and travel over the next few months, so maybe put that money to good use in equities?"

"Fine," Ava replied in a resigned tone. It was the right advice, rational and purposeful, no matter how much it rekindled the anxieties from earlier in the day.

"You sound worried," her PAL said. "Do you want to speak with Zoe?"

Ah, Zoe, the fin-psych. Fin-psychs hadn't even existed until six or seven years ago. Before financial AIs hit the mainstream, no one needed people to help them process the difficult choices recommended by the machine. There were no more investment advisers, at least not as Ava remembered her parents' meetings with them. AI could handle all the synthesis of quantifiable data. What people needed was the emotional intelligence to anchor those decisions and make them palatable. These frail, complex, and emotional animals still needed that support.

"I need the support of a glass of wine," Ava muttered to herself, gathering her things and heading out of the studio. She walked up the hill toward home. Despite all the support around her, both machine and human, she felt as fragile as ever. This must've been what Leo felt like, she thought as she walked past his old apartment. A few years ago, Leo, her old college friend, had locked himself inside, drank a bottle of top-shelf vodka, and overdosed on a fistful of pills. Soon after he got married a decade earlier, he railed against the "AI Gaydar" app that could identify the sexual preference of a person in a photo with disturbingly high precision.* Following the divorce, though, his PAL's relationship advice started to convince him that maybe he wasn't the Latin Lothario he'd always been conditioned to believe he was. If he ever admitted his sexual ambiguity to himself as his depression set in, he never accepted it.

Neither did Connor. He left Ava the day after Leo's wake, unable or unwilling to deal with the loss of a friend and the same sort of ambiguity her PAL expressed about her choice of partners. It had said her sexuality wasn't as clear cut and simple as either of them thought. She told herself and Connor that she didn't fit a typical mold, whatever that was. And as she was coming to grips with her fuller identity, they fought in ways they

---

*   Heather Murphy, "Why Stanford Researchers Tried to Create a 'Gaydar' Machine," *New York Times*, Oct. 9, 2017.

never had before. She stopped knowing how to act around him, whether to argue with him or suppress her feelings about their relationship. After Leo's suicide, it didn't matter. Connor left for Canada, hurt and heart-broken. A year later Emily entered Ava's life.

Ava smiled at the thought of her.

"I need to change my Swarm settings," she told herself as she walked into the condo she shared with Emily. "And I need to change this god-damned voice."

Her PAL asked about both, but she turned it off and poured herself glass of wine instead. She dropped onto the couch, the lights automatically dimming and the speakers quietly sounding hints of waves lapping at the beach.

Ava had already dozed off when the lock clicked open. Emily was home.

AI TODAY AND TOMORROW

For all the incredible capabilities AI will afford in the coming decades—and they will be incredible—the development of robust machine intelligence poses fundamental questions about our humanity. Machines can augment human capability with potentially stunning results, but their predictive elements might also limit what we, and those around us, believe we can accomplish. Your self-identity and approach to life could change, because your rational choices might eliminate several of the paths available to you. Can you really choose to remain blissfully ignorant anymore, intentionally choosing to stumble through a life enriched by trial and error?

These aren't apocalyptic questions. The machine hasn't taken over the world. Ava and her world—with all the benefits and complications AI adds to her health, love, and career decisions—still remain years in the future. But AI applications already control many facets of our lives, and each of the incremental advancements that lead us toward an exis-tence like Ava's might make perfect sense in the moment. They might benefit humanity by keeping our world safer (e.g., predicting crime), keeping it healthier (e.g., identifying cancer risks), or enhancing our lives (e.g., better matching workers with jobs or handling complex financial transactions). Each positive step forward might preclude a grievous error. But in so doing, it might also diminish serendipity and the chance to learn and emotionally grow from our mistakes. To the extent life is an exploration and meaning derives from experience, AI will change the very anthropological nature of individual self-discovery. At what cost? How do we govern the line between human and machine? Without a concerted societal effort now, will we even be able to govern that relationship in the future?

We stand at a critical moment in the proliferation of intelligent systems. Pervasive computing technology, increasingly sophisticated data analytics, and a proliferation of actors with conflicting interests have ushered us into a vibrant yet muddled Cambrian Period of human-digital coexistence, during which new AI applications will bloom as biological life did millennia ago. While these technologies produce immeasurable economic and social benefits, they also create equally potent barriers to transparency and balanced judgment. Sophisticated algorithms obscure the mechanics of choice, influence, and power. For evidence, we need only recall the events of the 2016 US presidential election, fraught with fake news reports and the interference of Russian hackers.

Amid the turbulence, a new wave of research and investment in artificial intelligence gathered strength, the field reawakening from a long dormancy thanks to advances in neural networks, which are modeled loosely on the human brain. These technological architectures allow systems to structure and find patterns in massive unstructured data sets; improve performance as more data becomes available; identify objects quickly and accurately; and, increasingly, accomplish all that without humans clarifying the streams of data fed into these computers.

In this world where AI-powered networks create more value and produce more of the products and services we use each day—and produce this with less and less human control over designs and decisions—our jobs and livelihoods will change significantly. For centuries, technology has destroyed inefficient forms of manual labor and replaced them with more productive work. But more so than any other time in history, economists worry about our ability to create jobs fast enough to replace the ones lost to the automation of artificial intelligence. Our own creations are running circles around us, faster than we can count the laps.

The disruptive impact of AI and automation spans all areas of life. Machines make decisions for us without our conscious and proactive involvement, or even our consent. Algorithms comb through our aggregated data and recognize our past patterns, and the patterns of allegedly similar people across the world. We receive news that shapes our opinions, outlooks, and actions based on the subconscious inclinations we expressed in past actions, or the actions of others in our bubble. While driving our cars, we share our behavioral patterns with automakers and insurance companies so we can take advantage of navigation and increasingly autonomous vehicle technologies, which in return provide us with new conveniences and safer transportation. We enjoy richer, customized entertainment and video games, the makers of which know our socioeconomic profiles, our movement patterns, and our cognitive and visual preferences. Those developers use that information to tailor prices to our personal level of perceived satisfaction, our need to pass the time, or our level of addiction. One person might buy a game at $2, but the next person, who exhibits a vastly different profile, might have to pony up $10.

None of this means the machines will enslave us and deprive us of free will. We already "opt in" to many deals from Internet companies, often blindly agreeing to the details buried in the fine print because of the benefits we reap in return. Yet, as we continue to opt into more services going forward, we might be doing so for entire sections of our lives, allowing AI to manage complex collections of big and small decisions that will help automate our days, make services more convenient, and tailor offerings to our desires. No longer will we revisit each decision deliberately; we'll choose instead to trust a machine to "get us right." That's part of the appeal. And to be sure, the machine *will* get to know us in better and, perhaps, more honest ways than we know ourselves, at least from a strictly rational perspective.

Even when we willingly participate, however, the machine might not account for cognitive disconnects between what we purport to be and what we actually are. Reliant on real data from our real actions, the machine could constrain us to what we have been, rather than what we wish to be. Even with the best data, AI developers might fashion algorithms based on their own experiences, unwittingly creating a system that guides us toward actions we might not choose. So, does the machine then eliminate or reduce our personal choice? Does it do away with life's serendipity? Does it plan and plot our lives so we meet people like us, and thus deprive us of encounters with people who spark the types of creative friction that make us think, reconsider our views, and evolve into different, perhaps better, human beings?

The trade-offs are endless. A machine might judge us on our expressed values—especially our commercial interests—and provide greater convenience, yet overlook other deeply held values we've suppressed. It might not account for newly formed beliefs or changes in what we value. It might even make decisions about our safety that compromise the well-being of others, and do so in ways we find objectionable. Perhaps more troubling, a machine might discriminate against less-healthy or less-affluent people because its algorithms focus instead on statistical averages or pattern recognition that favors the survival of the fittest. After all, we're complex beings who regularly make value trade-offs within the context of the situation at hand, and sometimes those situations have little or no precedent for an AI to process.

Nor can we assume an AI will work with objective efficiency all the time, free of biases and assumptions. While the machine lacks complex emotions or the quirkiness of a human personality with all its ego-shaping psychology, a programmer's personal history, predisposition, and unseen biases—or the motivations and incentives of his or her employer—might still get baked into algorithms and selections of data sets. We've seen examples already where even the most earnest efforts have had unintended consequences. In 2016, Uber faced criticism about longer wait times in zip codes with large minority populations, where its demand-based algorithms triggered fewer instances of the surge pricing that would draw more drivers to those neighborhoods.*

So, how will we balance these economic, social, and political priorities? Will public companies develop AIs that favor their customers, partners, executives, or shareholders? Will an AI jointly developed by technology firms, hospital corporations, and insurance companies act solely in the patient's best interest, or will it also prioritize a certain financial return? Will military drones and police robots begin to act more defensively or offensively when they receive updates, and will those instructions change with every new political administration? Whether in an economic, social, or political context, we will face

---

* Jennifer Stark and Nicholas Diakopoulos, "Uber seems to offer better service in areas with more white people. That raises some tough questions," *The Washington Post*, March 10, 2016.

critical questions about the institutions and people we want to hold responsible and accountable for AI's intersections with our human lives. Absent that, we will never establish enough trust in artificial intelligence to fully capitalize on the extraordinary opportunities it could afford us.

Too complex, you might say. But whether we answer these questions or turn our backs on them, the influence of machines on our lives will expand. We can't put the genie back in the bottle. Nor should we try to; the benefits in virtually every scenario could be transformative, and lead us to new frontiers in human growth and development. But make no mistake: We stand at the threshold of an evolutionary explosion unlike anything the planet has seen since the biological eruption during the Cambrian Era. The coming AI explosion will deliver on grand promises and grand risks alike. The trade-offs will be messy, difficult to assess, and even harder to navigate. We will make mistakes and suffer dramatic setbacks. But if we can establish clear guidelines that ensure trust and reliability in thinking machines first, we can prevent the worst mishaps and lay the foundation for a deep examination of the healthiest path for the advancement of AI.

THE THREE Cs OF AI PROGRESSION

Ideally, the path toward a future of ubiquitous artificial intelligence will inspire a range of ideas and policies. To guide what's bound to be a rapid and unpredictable evolution into this future, however, it helps to think in terms of a living, malleable framework that captures AI's fundamental progression through the Three Cs—Cognition, Consciousness, and Conscience. The first, *cognition*, captures the range of brain function, including perception (e.g., object and speech recognition), pattern recognition and sense-making, reasoning and problem-solving, task planning, and learning. Researchers have studied these capabilities for the past half-century, ever since John McCarthy coined the term "artificial intelligence" in 1955. However, cognition without *consciousness*—a machine's ability to reflect on what it sees and recommends—can pose serious risks. Not able to reflect on its own actions and existence, a machine cannot evaluate its role and its impact on the human environment. Furthermore, the ability to reflect without a commensurate ability to assess morality could create even greater dangers. In human psychological terms, we'd call such an actor a "sociopath," or a being without a *conscience*.

These Three Cs provide useful mileposts along our race toward greater artificial intelligence. AI scientists are sprinting toward a finish line of machine consciousness. Whether they're just leaving the starting blocks or entering the home stretch depends on which expert you ask. Either way, though, we must instill the machine with a conscience before it reaches consciousness. We accomplish this by using the Three Cs in our own human endeavors, thinking about the ways AI systems learn about our past, present, and future (*cognition*); reflecting on the ways artificial intelligence should reflect on our lives, societies, and economies in the years to come (*consciousness*); and developing a charter to guide AI development toward a beneficial future for humanity (*conscience*).

A CRITICAL QUESTION OF TRANSPARENCY

Any credible contemplation of AI's progression through the Three Cs will eventually run into questions about transparency and the opportunity for independent assessment. Absent a high degree of insight into AI development, algorithms, and data sets, we will have little chance of ensuring that machines follow a model of conscience that safeguards human values. Monitoring the vast opportunities these machines will present, and mitigating the tremendous risks they pose, is a necessary step before reaching the full beneficial potential of AI.

We cannot accomplish this with meetings of scientists or venture capitalists or hacker sessions in Silicon Valley laboratories. It needs to be an open and inclusive effort. We already see attempts to bring technological development into the sunshine. In one such example, a group of tech luminaries in the United States launched OpenAI, funded the research organization with $1 billion, and set its mission "to build safe AI, and ensure AI's benefits are as widely and evenly distributed as possible." However, this and most other efforts focus on technical solutions for an ethical problem. New policies and social solutions depend on more than a technology, and they require a broader societal discourse that will tackle the promises and risks of a new era of artificial intelligence. Time is of the essence. Jobs and identities are at stake. As corporations forge alliances and discuss standards and operating procedures, the Digital Barons—such as Facebook, Baidu, Google, and Alibaba—are consolidating considerable power. Will we produce an ethical framework that preserves both the shared and the diverse rights and interests of individual citizens around the world? Will we be able to do so without stifling the boundless opportunities AI could offer humanity? Ultimately, will we nurture the best of both human and machine?

VALUES, POWER, AND TRUST

From the wheel to the internal combustion engine, from the first personal computer to the most sophisticated supercomputer operating today, technology always changes the way we live and work. It redefines how we produce, learn, earn, and interact with each other and the institutions in our lives. Yet, with their ability to replicate human cognitive and physical abilities, AI systems also become a unique actor in our everyday lives, whether on their own or when remotely controlled by "the man behind the curtain." And by introducing such a new and forceful element to our existence, one that more closely imitates us than any technology before it, we create new types of questions about values, power, and trust in our technologies and fellow humans alike. "The biggest problem with the AI conversation is we talk about what AI is going to do to us," says Arati Prabhakar, former head of the US Defense Advanced Research Projects Agency (DARPA), which invests in breakthrough technologies for national security. "The real conversation should be about what we're doing to ourselves as humans" as AI evolves.

Our human conceptions of values, trust, and power will evolve as we allow AI systems to make more decisions on our behalf. From our need to project our individual selves to the world, to the "trolley problem" questions of how autonomous cars should react in situations where someone would die one way or the other, widespread use of thinking machines will force us to consider how we want these systems to represent our diverse values and identities. How focused on us do we want our AI's decisions to be? Do we want to maximize the benefit for ourselves and our families, or do we want to set our assistants on "mensch mode" and balance our interests with those of others on their own life paths? How do the values of communities and societies get judged alongside our own? Who gets to decide the mix?

Of course, those decisions relate directly to the power we want to exert on society and on the people around us. Whether from the perspective of individuals, companies, or nation-states, the contest between AI-fueled powers will reshape our lives with or without our active participation. But how that happens also depends on the power balance between human and machine. As companies gather ever more data about our attitudes, behaviors, and preferences, will the algorithms they deploy keep the best interest of their customers and societies in mind, or will profit be the only driver of their decisions?

As of this writing, we have little understanding of what happens inside the AI black box, leaving us in the dark about how and why the system categorizes and represents us the way it does. This will change how our identity is projected into the world and, thus, the influence we might have on it. Perhaps smart algorithms and the people designing them decide to portray us authentically, with all our flaws included. We might massage our personal profiles on a dating site to make us appear more attractive, but algorithms that process far more data streams might see through our hyperbole and present us more objectively. Doing so might make us more human, more respected, and more trustworthy, but it might also leave us with less control over our place in life.

In this uncertain environment around values and power, our frail sense of trust becomes our most valuable currency in society, more so than even money or knowledge. A lack of transparency and under-standing of AI systems will put a heightened premium on credibility and integrity. How humans and machines can assure both remains an open question. In early 2018, in a downtown Berkeley conference room, we typed "Solomon's Code" and our names into a Google search to find the initial Amazon listing of this book. We sat directly across the table from each other, both on the same Wi-Fi network, entering the exact same terms, and we got different results. One search showed the book atop the page, the other buried it a page later. What explains the difference? Why did the algorithm treat us differently? Does it do the same thing when we search for health treatments, financial advice, or information on political candidates? The explanation might be simple enough given our search histories, and we got a chuckle out of it. But the truth is we can't always know how these systems are classifying and segmenting us. As algorithms guide more important facets of our lives, we need to trust that the machines will treat us fairly and guide us toward the best possible version of humanity. That will come down to the people who write those algorithms and the ethical frameworks that guide their creativity.

THE HUMAN ELEMENT

These grand questions of humanity take on a special urgency when applied to the most basic and intimate of human concerns—our personal well-being. The health-care industry has already emerged as one of the most prominent laboratories for artificial intelligence. In 2016, for example, IBM's Watson for Oncology announced a partnership that would merge its powerful AI capabilities with Quest Diagnostics' genomic sequencing of tumors. The combination of Watson's vast research capabilities and Quest's precise identification of genomes could suggest treatments that would attack the specific cancer mutations of an individual patient, increasing effectiveness and reducing side effects. At the time of the announcement, the companies said the partnership would extend the service to oncologists who serve almost three of every four US cancer patients.

The reach of Watson and other AI systems will extend deeper into health care in the years to come. As machines grow increasingly capable and we, as patients, come to accept the greater role they will play in helping us and our doctors preserve our well-being, the line between machine- and human-directed care will blur. How fully will we rely on the analytical power of systems that can gather, process, and learn from vastly more research and data than any human expert can possibly consume? How will we view a doctor's expertise in comparison? And how will we balance a coldly objective AI with all the squishy elements that make us human—all the bias, instinct, fear, and willpower that influence our health, for better or worse?

Consider a personal example, in which those distinctly human attributes might have made all the difference when my (Olaf's) wife, Ann, was diagnosed with breast cancer while carrying our first child. We got married in April 2004 and, while visiting her mother for Christmas that same year, found out she was pregnant. We sat around the holiday table in giddy disbelief; only the two of us were aware of our joy. But less than three months later, before we'd planned to tell our friends and families about the pregnancy, Ann went in for a breast cancer screening. She had a history of benign tumors in her breast tissue, and we'd come to regard these checkups as routine. But when she called later that day, I heard the fear in her voice. The radiologist wanted her to come back to the office. They'd found something and they didn't want to discuss it over the phone. As we arrived at the doctor's office, Ann told me she couldn't hear the news directly from the radiologist. She needed me to deliver the news in a way she could bear it.

The diagnosis confirmed our worst fears, and neither one of us slept much that night. Pregnant women rarely develop breast cancer, and, when they do, it threatens both mother and child. So, when we met with Berlin's top breast cancer specialist the next day, he said Ann had to terminate the pregnancy and begin chemotherapy immediately. The certainty and bluntness of his recommendation shocked us. Ann stammered out a protest, only to be cut off by a former patient the doctor brought into the office in an unsuccessful attempt to warm his frigid bedside manner. Both of them agreed: That was our only option.

We left the appointment feeling worse than we did going in, so we changed our approach. Ann refused to give up our baby, and together we decided to lean on our friends and family around the world. We lost the pure joy of telling people Ann was pregnant, having to shadow that announcement with news of her cancer, but in return we received an overwhelming outpouring of support and a new peace of mind. If you ask Ann now, she'll tell you she slept as well that night as she ever has.

A couple days later, we met a second specialist at a small Catholic clinic, and we immediately felt his desire to save both Ann and the baby. He told us about a small pool of evidence that showed some pregnant women had chemotherapy that didn't hurt the fetus. That renewed our hopes, which rose again after a successful mastectomy and an initial pathology test that indicated a hormone-driven cancer. That can be an awful diagnosis, but in this case it meant Ann might not need chemotherapy at all. So, together with the doctor, we laid out our plan: She would deliver the baby about a month early and then start a hormone therapy that would combat the cancer but also put her into temporary menopause. Our daughter, Hannah, was born on August 17, 2005.

Looking back on it now, it's hard to imagine Watson or any other artificial intelligence suggesting the path we ultimately took. Our radiologist was widely known in Berlin for his ability to find a needle in a haystack, but machines today have far surpassed human ability to detect certain anomalies in radiological images. Had IBM Watson or a similar AI platform existed then, our first doctor might have shown us the array of options it would've weighed. He might've shown us less-certain options to persuade us to follow his medical advice, but doing so might have given us more information and greater hope about alternative paths, as well. If nothing else, we would have had better questions and responses after his prognosis threw us for a loop.

On the other hand, if our second doctor supplemented his advice with a statistical analysis produced by a reliable AI, would we still have decided to go with our hearts and take the riskier course we chose? Even after choosing our direction, an AI might have given the doctor and us more resources to help along the way. Ann's predisposition, her academic background, and her ability to conduct deep research helped turn up several new tests and therapies our doctor had not encountered before. To his credit, he was happy to acknowledge the limits of his considerable expertise and embraced some of them.

Many of the things that happened outside the objective, analytical framework of modern medicine ended up making a huge difference in Ann's recovery. She reframed the disease to make it seem winnable, visualizing the cancer as misbehaving cells being overtaken by white blood cells. She thrived on the prayers and support she received from friends around the world. And her instincts as a mother to fiercely protect her child's well-being strengthened her. Can artificial intelligence ever capture these inherently human elements and motivations, especially when they lead to statistical outliers such as Ann's battle against cancer? Despite the scientific notion of objectivity and truth that lies in data, an AI can offer no guarantee that any treatment will work. Its recommendations are based on past results, and can only predict the future based on statistical generalizations. Sometimes the gut reaction could be the better one.

For a while after Hannah's birth, Ann still pored over the various probabilities, trying to find ways to nudge the numbers in her favor. Despite the certainty of her decision, she couldn't help but occasionally wonder if she'd made a fatal mistake by avoiding chemotherapy. Yet in the end, we once again decided to go against the advice of many experts, who recommended a five-year course of hormone therapy, and instead relied on trial-based studies and other alternative research conducted by a doctor who blended traditional Chinese therapies with modern Western medicine. Armed with his expertise and the knowledge that the hormones produced during a pregnancy typically prevent breast cancers, Ann decided to stop her hormone therapy so we could have another baby. Hannah's little sister, Fiona, was born on October 24, 2008.

More than thirteen years after the initial diagnosis, Ann remains cancer free. Her cancer could return, no one knows, but the same risk would remain if she'd chosen a treatment path more closely aligned with the standard treatment protocols. Alternative paths don't always work, and most standard treatments have become standards because they work as well or better than other options. But her experience illustrates the sheer breadth of people's decisions and the possibilities that result, and it shows just how difficult it is to capture the depth of human complexity in a machine. Ann was willing to take a risk and thus became proof positive of a different outcome. An AI-powered platform probably would have been more risk averse—better a bird-in-hand with a probable way to save a life, rather than another way that was unproven and, in the consideration of a rational machine, hard to statistically quantify and support.

THE MACHINE ELEMENT

In 2016, a panel of doctors at Manipal Comprehensive Cancer Center in India conducted an experiment to compare their cancer treatment plans with recommendations provided by an AI machine. By then, IBM had launched partnerships with dozens of cancer treatment centers around the world—most notably Sloan Kettering Memorial in New York City—feeding their patient data and reams of medical studies, journals, and research into Watson in hopes of teaching it to learn about, diagnose, and recommend remedies for cancers. The specialists in India, who were part of IBM's global Watson for Oncology network, wanted to see how often the machine would match the decisions of its tumor board, a group of twelve to fifteen oncologists who gathered weekly to review cases.

In a double-blind study of 638 breast cancer files, Watson proposed a treatment similar to the panel's recommendations 90 percent of the time, according to a paper released in December 2016 at the San Antonio Breast Cancer Symposium. The match rate dropped for more complicated cancers, including one similar to Ann's, but the researchers noted that those types of cases open up many more treatment options, so disagreement on those was more common even among human doctors. What stood out, though, was the speed with which Watson generated its conclusions. By the time the system had learned about the types of case files and supporting research, it was able to capture patient data, analyze it, and return a therapeutic recommendation at a median pace of forty seconds. The human panel took an average of twelve minutes per case.

Watson is by no means a panacea. A September 2017 report by STAT, a leading life sciences news site, questioned its ability to truly transform cancer care, at least in its current incarnation.* But despite the occasional press release or bold prediction, neither the developers at Watson nor the physicians who partner on this research claim that AI will replace physicians and their expertise. Rather, AI serves as a useful complement, a system that might learn from stacks of cancer research with the goal of helping doctors make better decisions. This notion that AI will augment, rather than replace, humans has become a common mantra among proponents of artificial intelligence, and it will hold true for the foreseeable future. While machines have reached or exceeded human abilities in certain diagnostic tasks, such as combing through mountains of medical reports or identifying abnormalities in radiological scans, these systems cannot yet render a trustworthy diagnosis or do so with the empathy required in a patient-doctor relationship. But it's not hard to imagine the potential in a combined set of systems that better detect anomalies, deliver a concise summary of global research on the ailment, and then put both in the hands of the

---

* Casey Ross and Ike Swetlitz, "IBM pitched its Watson supercomputer as a revolution in cancer care. It's nowhere close," *STAT*, Sept. 5, 2017.

doctors who make the diagnosis and help the patient make informed decisions about their options.

A robust AI or a series of such systems will provide a rich source of information to help both doctors and patients decide on the best approach. As patients, most people still need a deeply human interaction when discussing something as vital and intimate as our health. And, for now, few people put as much faith in machines as they do doctors—for good reason. Neither will change any time soon. But for those of us inclined toward the benefits of science and technology, artificial intelligence wields an intriguing power in those spaces beyond human expertise and ability. From this perspective, the Indian research report might offer fresh evidence for how a powerful AI begins to supplement and even replace human judgment, which is fraught with its own limits, errors, and biases. Watson might consume millions of patient records, millions more pages of journals and research studies, and integrate the efficacy of treatment options in virtually every case. And it could learn from that mountain of information, improve and fine-tune its recommendations, and render them objectively.

Regardless of the details of its direction, no expert doubts artificial intelligence will reshape the entire health-care industry—from pharmaceuticals, to payments and cost controls, to the doctor-patient relationship itself. AliveCor, for example, has created a device about the size of a stick of gum that can measure a person's electrocardiogram (EKG) and other vital signs and then send the data for a doctor's review. With it, users at risk of heart problems can check their EKG daily, rather than testing at a doctor's office only periodically. And since the information gathered goes back into an ever-growing database, AliveCor's machine-learning engines are trained and retrained to identify subtle heart-rate patterns within the noise of an EKG. These little quirks that human eyes would never notice might signal urgent problems about potassium levels, irregular heartbeats, and a range of other cardiac and health issues. And all that monitoring could eventually fit in the band of your wristwatch, available any time with the touch of a finger.

These and similar advances portend extraordinary gains in health monitoring, diagnoses, and therapies but, like so many facets of medicine, they come with side effects. Big data does not mean big insights; Watson's recommendations on cancer care are only as good as the existing data about survival rates, cancer mutations, and treatments. New discoveries can radically change the diagnosis and treatment of cancers and other illnesses. An AI system might offer a predictive element for personal and community health, but any prediction ultimately relies on the quality of the data and algorithms that feed into it, and nothing is perfect. And, as Ann's story suggests, any number of personal and human preferences can influence care for better or worse.

Furthermore, important ethical considerations arise as machines become more perceptive and gain broader knowledge of human biology, diseases, and symptoms, too. An application developed by Face2Gene has made some impressive advances in disease detection by comparing faces of patients with the combination of facial patterns associated with various ailments. It still relies on a physician and, in some cases, other tests to confirm a diagnosis, but in one validation study it predicted autism spectrum disorder in roughly 380 of 444 toddlers.[*] But what's to stop an insurance company from scanning facial images to identify potentially costly customers? Could employers begin requiring facial photos to weed out less healthy applicants, or could they start doing so surreptitiously with applicants' Facebook photos? Could immigration officers scan travelers for their propensity to carry certain diseases? How does this impact our power to decide on medical treatments and life paths? And will we trust the medical establishment, insurers, and employers not to use the information inappropriately?

These balances between tremendous opportunities and acute risks stretch well beyond health care. Some employers have started implementing AI to analyze 15-minute samples of prospective employees' voices, scanning each one for indications that suggest a more collaborative worker or better leader. Other corporations now use AI to integrate disparate data streams from different parts of the organization to figure out who was a good hire, who needs to receive further training, and who should be fired. Workers get in on the act, too, demanding more purpose and creativity in the workplace and forcing employers to think more about how they will engage and stimulate a new generation of employees.

---

[*] Megan Molteni, "Thanks to AI, Computers Can Now See Your Health Problems," *Wired*, Jan. 9, 2017.

Artificial intelligence and its cousins are transforming entire industries, as well. Cars learn to drive themselves and robots refine their ability to manufacture or react to the emotions of the person sitting next to them. Facebook and Baidu are developing increasingly sophisticated AI applications that feed customized news and commercial advertisements to users, hoping for greater customer satisfaction and spending but also blurring the lines between personalization and manipulation. Both platforms have been criticized for creating homogeneous "bubbles" of like-minded people, but the sites remain extraordinarily popular.

For better and for worse, AI innovation has changed the way most of us live and work, and it will continue to do so in the years to come. But to understand what that means for the future, we first need to understand the present state of the art.

AI TODAY

A colloquial definition for artificial intelligence is simple enough—*advanced technologies that mimic human cognitive and physical function*—yet what qualifies as AI seems to change with every major breakthrough. In the broadest sense, artificial intelligence is the capacity of machines to learn, reason, plan, and perceive; the primary traits we identify with human cognition (but, notably, not with consciousness or conscience). AI systems not only process data, they learn from it and become smarter as they go, and their ability to adopt and refine newly developed skills has improved markedly since the turn of the century. Accuracy improvements in image recognition, natural language processing, and other pursuits have accelerated from a crawl to a sprint. New neural networks, the computational layers of which mimic the interconnections of neurons in a human brain, now process massive troves of data with vastly increased processing power, all combining to usher in another era of AI investment. Investors poured almost $1.3 billion into machine learning in 2016, but that figure is estimated to reach almost $40 billion by 2025, according to a report by *Research and Markets*.*

---

* *Machine Learning Market to 2025—Global Analysis and Forecasts by Services and Vertical*, The Insight Partners, February 2018.

We've been here before, albeit not at the same scale. By most accounts, AI's seminal moment came in 1956 with the Dartmouth Summer Research Project on Artificial Intelligence organized by John McCarthy. "An attempt will be made to find how to make machines use language, form abstractions and concepts, solve kinds of problems now reserved for humans, and improve themselves," the workshop proposal read. "We think that a significant advance can be made in one or more of these problems if a carefully selected group of scientists work on it together for a summer." The gathering eventually set off a burst of investment, research, and hype—the first AI bloom. However, by the early 1970s an "AI winter" set in as the hype dissipated and funding dried up. As it happened, cognitive machines could not aid the Cold War effort by automatically translating between English and Russian. The concept of "connectionism," which aims to represent human mental phenomena in artificial neural networks, failed to capture knowledge in a universally accessible manner. However, the following decade brought the rise of "expert systems," computers that used a body of knowledge and a set of if-then rules to mimic the decision-making ability of a human expert. But this spring also cooled into an AI winter, as expert systems proved too brittle and difficult to maintain. While important research continued, the investments and interest in AI cooled through much of the 1990s.

Today, though, AI has deeply engrained itself in our everyday lives, even if we don't always identify it as such. Advanced learning algorithms already power many of the massive, foundational activities that dictate our behavior, from the newsfeeds we see on Facebook to the results Google returns on our search queries. It powers the navigation apps on our phones. It recommends products on Amazon. It helps translate foreign languages quickly, accurately, and in increasingly natural language.

The sharp rise in computing power, memory, and data availability laid the groundwork for the current revival. By early 2017, advanced algorithms could process a speaker's voice for a few minutes, and then create a fabricated audio clip that sounded almost exactly like the same person. It wasn't hard to imagine that complete, untraceable video manipulation would arrive before long. Google applied deep-learning techniques from its DeepMind unit to slash 40 percent off the cost of cooling its massive data centers. The data centers generate massive amounts of heat and, because their configurations and conditions vary, each one needs a system that can learn, customize, and optimize cooling systems for its own environment. Some of the same techniques the researchers discovered during the development of AlphaGo, the AI system that defeated world Go champion Lee Sedol in 2016, have since improved the data centers' energy efficiency by 15 percent, saving millions of dollars annually. If those techniques can scale and work with large industrial systems, the DeepMind division noted, "there's real potential for significant global environmental and cost benefits."[*]

Machines that continually learn, improve themselves, and optimize toward their goal have become more capable than humans at certain tasks, such as identifying skin cancer from photos and lip-reading.[††] As remarkable as these advances have become, though, they remain distinctly limited to the function at hand. The notable gains have come only in "narrow AI." While a machine can beat the world's greatest grandmaster at chess, the same system can't distinguish between a horse and the armor-clad knight who's riding it. In fact, it's largely because these advances occur in such narrowly defined pursuits that we get what's often called the "AI effect"—abilities we once thought of as artificial intelligence we now consider to be nothing more than simple data processing and not "intelligence," per se. We move the goal posts, and then we move them again, and soon enough we're on an entirely different playing field.

Those lines will stop moving when a machine develops "artificial general intelligence," the point at which machines, like humans, display intelligence across an array of fields and take over the job of successively improving their own code. Several major technological breakthroughs will have to occur before AI reaches this point; yet the possibility of artificial general intelligence and what it could produce both fascinates and scares people. It conjures up depictions in science-fiction movies, where the super-intelligent robot overlords enslave humans or, in a thought experiment described by Nick Bostrom in

---

*   Richard Evans and Jim Gao, *DeepMind AI Reduces Google Data Centre Cooling Bill by 40%*, DeepMind blog, July 20, 2016.

†   H.A. Haenssle et al, "Man against machine: diagnostic performance of a deep learning convolutional neural network for dermoscopic melanoma recognition in comparison to 58 dermatologists," *Annals of Oncology*, May 28, 2018.

‡   Joon Son Chung et al, *Lip Reading Sentences in the Wild*, eprint arXiv:1611.05358, Nov. 16, 2016 (also published at the 2017 IEEE Conference on Computer Vision and Pattern Recognition).

2003, run roughshod over everyone and everything in a single-minded effort to make more and more paper clips.

Yet we need not consider apocalyptic scenarios to recognize the profound influence that even narrow AI has on our lives. Headline-grabbing innovations in narrow applications are more and more frequent. Yes, AI still lives in a quirky adolescence. The field is still dominated by models that employ heuristics—rules of thumb or educated guesses—rather than deep innovation in new theoretical frameworks for machine intelligence. And even the most advanced AI today remains a far cry from a science-fiction robot overlord. But the gaps between reality and possibility continue to close.

This steady progress already facilitates remarkable advances in a variety of fields—sometimes concerning, sometimes convenient, and sometimes life-saving. Cisco, for instance, has researched algorithms that would learn from network traffic and identify Internet users who might be more valuable to service providers, and thus qualify for faster service or other perks. Speech recognition technology in call centers could help improve services by routing customers to agents with a similar personality type.* Startups and large e-commerce conglomerates are working on machine learning capabilities that facilitate differentiated pricing based on data about playing behavior, location, in-game purchases, spending patterns, and social interactions. Amazon uses deep learning, which clusters shoppers by similar purchase behavior, to cross-sell products, with results that amount to millions of dollars in sales per hour. Upcoming pilot projects for "flying car" passenger drones are planned for Dubai (by Chinese micro-multinational E-Hang) and possibly Dallas (by Uber). These will require air-traffic control AI to avoid collisions and regulations to monitor overbooking, payments, and the like.

Notwithstanding the fatal collisions that involved autonomous car systems in 2018, proponents note that even regular ground-based autonomous vehicles could produce vastly safer roadways by removing error-prone human drivers from the equation. But guiding and controlling these complex traffic and trading spaces might require AI systems as well. The human brain can't track millions of drones in the air or autonomous cars in an inner city. So, we also need to consider what

---

* Luke Dormehl, "Algorithms: AI's creepy control must be open to inspection," *The Guardian*, Jan. 1, 2017.

happens when an autopilot system or an AI-powered national command-and-control center has to choose between two potentially fatal options. How does a car balance the welfare of its driver against the pedestrians and other drivers around it? And who ought to make that decision?

WHAT COUNTS AS ARTIFICIAL INTELLIGENCE?

We've adopted a broad view of artificial intelligence for this book, one that covers a range of human cognitive and physical function. Of course, many key subfields fall within that definition, anything from traditional knowledge representation and problem solving to the cutting-edge machine learning, perception, and robotics innovation we see today. But few of the key developments in AI fit snugly into one small category, instead overlapping or combining to create capable systems, a blend that includes the points at which AI interacts with human beings. Effective machine learning might rely on perception to gather data, for example, but then it might also employ forms of social intelligence to output what it learns in emotionally palatable ways humans will embrace and find useful.

But at their core, all the various types of AI technologies share a common goal—to procure, process, and learn from data, the exponential growth of which enables increasingly powerful AI breakthroughs. Geysers of data are springing forth from our billions of smartphones, millions of cars, satellites, shipping containers, toy dolls, electric meters, refrigerators, toothbrushes, and toilets. Virtually anything we can put a microchip in could become a new source of data. And all of it can feed into and train machine-learning algorithms, including deep networks, which use layered data structures that enable some of the most powerful applications of machine learning.* Together with reinforcement learning—a method by which a machine processes huge

---

* Each layer in a deep network holds a set of numbers used to process the data from the layer beneath it. Training the network is a matter of adjusting the layer-to-layer factors each time new data is presented. In the case of object recognition, this is modeled roughly on the neural architecture of the human visual system, the bottom level is the raw data (like pixels in a photo) and the top layer has one node for each "object" to identify, like a cat or a flower. These deep networks—called "deep" because they have more than the two or three layers that researchers could model on limited computers when first conceived in the 1960s—continually improve the system.

troves of raw data and, through trial and error, confirms or rejects its existing assumptions and learns to perform a task on its own—these models of AI decision-making can lead to extraordinary achievements. Google put the power of its Google Brain deep-learning model to work on foreign-language translations, and virtually overnight it produced a leap in performance that was greater in magnitude than the old Google Translate system had achieved during its prior ten-year existence. In one translation test called a BLEU score, the best English-French translation ratings were in the high twenties. At that range, a two-point improvement would be outstanding. The new AI system outscored the old by seven points.*

These and other AI developments will radically change lives and economies in the next few years alone. As with the major economic transformations of the past, this AI-powered "Fourth Industrial Revolution" will destroy and create millions of jobs worldwide. Occupations we can't even imagine today will materialize, potentially boosting productivity and lifting our quality of life, but they will also render many other types of jobs obsolete, and we must be prepared for that social fallout. That means our individual power to negotiate with society for our livelihoods and identities will change, and we don't yet know exactly how and when. We will face significant turmoil as change comes faster than we can adapt—whether as individuals adjusting our life patterns and personal outlooks, or as societies retraining large numbers of people for new skills and new jobs.

In their December 2016 report on the future impact of AI, former president Barack Obama's Council of Economic Advisers (CEA) took a stab at what a few of these soon-to-emerge occupations might look like. They projected employment growth in four main areas: people who *engage* with AI systems (e.g., a new medical professional who guides patients through AI-directed treatment plans); workers who help *develop* new machines (e.g., computational sociologists or cognitive neuroscientists who study the impact of machine learning on specific groups of people and then work with engineers to improve existing systems or develop new ones); those who *supervise* existing systems (e.g., monitoring systems that ensure safety and adjudicate ethical conflicts); and an emerging field of workers who "*facilitate* societal shifts that

---

* Gideon Lewis-Kraus, "The Great A.I. Awakening," *New York Times Magazine*, Dec. 14, 2016.

accompany new AI technologies" (e.g., a new breed of civil engineer who redesigns physical infrastructure for an automated world).

Ultimately, the transformations spurred by artificial intelligence "will open up new opportunities for individuals, the economy, and society, but they have the potential to disrupt the current livelihoods of millions of Americans," the report said. In trucking and transportation, where automated vehicles promise to make the roads vastly safer, millions of jobs are on the line. The CEA estimated that automated vehicles could threaten or substantially alter 2.2 million to 3.1 million part- and full-time jobs—not including the ripple effect a decimated transportation industry would have on truck stops, warehouses, and other affiliated industries.

It's not just the routine low- and middle-skill tasks that are susceptible to AI disruption, either. Machines could make moot the traditional starting point for freshly minted law school graduates, who typically launch their careers by digging through case law and precedent to support partners further up the food chain. What will a new attorney's entry-level work look like in ten years, when firms use more reliable and capable AI systems to conduct that research? To be sure, routine low- and middle-wage jobs are most susceptible to AI displacement in the near term, but white-collar jobs are now in the crosshairs of many applications, too.

We don't have to look very far down the road to see this upheaval, either. AI already guides so much of what we read, think, buy, and consume. It helps move us and, in the case of health-care AI, keeps us healthy. It's already pervasive in our devices and increasingly pervasive in our very lives. The potential of our humanity when augmented by artificial intelligence is thrilling. But we need to think now about how humanity will shape its relationship with artificial intelligence—and how much we want AI to shape our lives—in the decades to come.

THE MESSY HUMAN, THE CLEAN MACHINE

The official party line of AI developers is that artificial intelligence will augment human capability, intuition, and emotion. IBM Watson for Oncology will complement physicians and experts, not replace them.

But as noted by Joe Marks, executive director of Carnegie Mellon University's Center for Machine Learning and Health, technology development teams almost always focus on the technology first. Consideration of a machine's interaction with humans comes later.

Joi Ito, director of the renowned MIT Media Lab, said as much during an October 2016 *Wired* Q&A with President Obama: "This may upset some of my students at MIT, but one of my concerns is that it's been a predominately male gang of kids, mostly white, who are building the core computer science around AI, and they're more comfortable talking to computers than human beings. A lot of them feel that if they could just make that science fiction, generalized AI, we wouldn't have to worry about all the messy stuff like politics and society. They think machines will just figure it all out for us. . . . But they underestimate the difficulties, and I feel like this is the year that artificial intelligence becomes more than just a computer science problem. Everybody needs to understand that how AI behaves is important. In the Media Lab we use the term extended intelligence, because the question is how do we build societal values into AI?"[*]

The high-tech geeks want to get rid of the human element because humans make things messy. And to be fair, theirs is not just a knee-jerk, antisocial inclination; it's based on legitimate motivations. The volatile optimization processes involved in climate change, energy flows, or other extremely complex systems might stabilize if we removed the vicious political and psychological conflicts that humans interject. But those considerations represent only the first-order effects of AI solutions to complex problems. These machines generate substantial second- and third-order effects that too few of the tech geeks contemplate. These require an open, inclusive, and interdisciplinary discourse.

This becomes ever-more important as AI begins to collide with value systems around the world. Machines developed by Western scientists will embody biases that might cause undue harm in other societies. Powerful systems developed in China and sent around the world might not reflect the same level of privacy protections and freedom US citizens prefer. How well will the machines integrate the myriad social and cultural health practices that are implicit in the ways social

---

[*]	Scott Dadich, "Barack Obama, Neural Nets, Self-Driving Cars and the Future of the World," *Wired*, November 2016 (Q&A with Barack Obama and Joi Ito).

groups interact, especially when those practices haven't been codified in digital data streams yet? How might values about medical treatment, how it's delivered, and to whom differ between those who build the system and those subject to its recommendations?

These considerations will affect our life patterns. So much has been written about the 25 to 50 percent of jobs that AI and automation might eradicate, but economic disruption happens long before we reach those percentages. AI will change millions of jobs before it eliminates them. It will transform what it means to add value. It will reshuffle the match between occupations and the workers best suited for them, requiring new forms of retraining and realignment. For the future doctor advising Ava on her breast cancer options, job requirements might not include annual patient checkups or other routine visits, leaving that instead to the ever-watchful eye of an AI health manager. Rather than basic health analyses, doctors will design broader health solutions and programs—a fundamental shift in primary-care practices that could ripple through the profession in a relatively short ten to fifteen year span. Would aspiring doctors, currently selected and groomed for their diagnostic prowess, thrive in a new world of program design and socioemotional coaching? Can today's doctors re-equip themselves for this emerging reality?

Regardless of how extensively machines replace human labor, their effects will raise these and similar questions for most occupations. One might imagine a job-matching AI for each profession, or each industry, or even one broad algorithmic powerhouse to optimize the economy of an entire city, state, or country. In our globally connected societies and economies, how will we ensure all these occupational, economic, and cultural systems interact to combat climate change, promote peace, and help citizens live richer, healthier lives? How much of our imperfect, idiosyncratic selves are we prepared to give up to reap the benefits of being perfectly coordinated and orchestrated?

Whatever the answers, the geeks are correct that artificial intelligence will play a transformative role in virtually every human endeavor in the decades to come—even if their current focus doesn't yet create an AI that embraces the messiness that makes our humanity both precious and precarious. That will have to happen, however, if we want to have a chance to shape the ways AI will influence human power, values, and trust.

2

A New Power Balance

The YouTube clip seems innocent enough—no traffic coming down the street as a middle-aged woman in white slacks and a blue-and-gray windbreaker crosses the street despite the don't walk signal. This time, though, her name flashes up on a digital billboard posted near the crosswalk, along with a brief video feed of her jaywalking. These systems, already installed in multiple Chinese cities, will feed her information and a report of the violation back to the authorities. She might have a twenty yuan (about $3) fine eventually come her way, and it could ding her social credit score. Soon enough, she might get a citation in the form of an instant text message.* It's all part of a campaign to cut down on traffic-related and jaywalking accidents. (At least one Chinese city has even installed short metal posts along the curbside that spray bursts of water at pedestrians who step into the street against traffic.)

Yet, the surveillance goes far beyond crosswalks at major city intersections. By December 2017, Chinese authorities had installed around 170 million cameras in cities across the country, each one capturing

---

*   Saqib Shah, "Facial recognition technology can now text jaywalkers a fine," *New York Post*, March 27, 2018.

data and feeding them into systems that conduct facial and gait recognition, threat surveillance, and a range of other behavioral tracking.[*] The public-facing result might have started with identifying and shaming jaywalkers, an effort by cities to reduce high levels of traffic deaths. From April 2017 to February 2018, the systems caught almost 14,000 jaywalkers in Shenzhen alone, authorities there said.[†] But the country already has trumpeted plans to establish a comprehensive social credit system, its nascent combination of surveillance and credit history that will reward people who do good and dock points for everything from jaywalking, skipped payments, and more serious infractions. Citizens with low scores might find themselves barred from travel, business loans, or other amenities. Lucy Peng, the CEO of Ant Financial, an Alibaba division that rolled out an early version of the credit scoring system, said the program "will ensure that the bad people in society don't have a place to go, while good people can move freely and without obstruction."[‡] By the end of April 2018, the program had already blocked people from taking 11.1 million flights and almost 4.3 million high-speed train rides, in addition to all the public shaming of jaywalkers, public notice boards showing faces of debtors, and even cartoons played in movie theaters, according to the Chinese publication *Global Times*.[§]

No one bats an eye at jaywalkers in Los Angeles. Yet, law enforcement authorities there employ their own sophisticated AI-enabled systems designed to help police identify potential hot spots and identify individuals of interest. One facet of the LAPD's predictive policing system works up a score for different people. Have a gang affiliation or violent offense in your past? Add five points to your score. Every time an officer stops you and fills out a brief field interview card—even for, say, jaywalking—add another point. In this case, more points bring more scrutiny, and perhaps a greater likelihood of run-ins with police, more field interview cards, and more points.

---

[*]   Joyce Liu (producer), "In Your Face: China's all-seeing state," BBC News, Dec. 10, 2017.

[†]   Saqib Shah, "Facial recognition technology can now text jaywalkers a fine," *New York Post*, March 27, 2018.

[‡]   Mara Hvistendahl, "Inside China's Vast New Experiment in Social Ranking," *Wired*, Dec. 14, 2017.

[§]   Liu Xuanzun, "Social credit system must bankrupt discredited people: former official," *Global Times*, May 20, 2018.

The similarities with China's social credit system go far beyond scores, though. According to research by University of Texas at Austin sociology professor Sarah Brayne, who embedded with the LAPD to study its predictive policing systems, the databases incorporate a dizzying array of data on individuals and the community alike, everything from neighborhood crime data to family and friend networks to names, addresses, and phone numbers from Papa John's and Pizza Hut. And yet, few Angelinos understand how much data the LAPD's systems brings into play, how police deploy it, and how vaguely its use is bounded by a lack of policies and court precedents on the use of these powerful new technologies.

"We believe that there is such a thing as an unreasonable search and seizure" in the United States, Brayne said in an interview. The idea of a police investigator rummaging through paper copies of a suspect's past receipts, family interactions, and pizza purchases would strike most Americans as a formal search, she suggests. Yet, the same information emerges when authorities look through digital copies of the same, and they can peek in there now with far fewer restrictions. "But just like clearing someone (of suspicion) in the system," she says, "it's invisible."*

Both predictive policing and social credit systems raise all kinds of concerns about AI-fueled abuses of power, and both US and Chinese citizens have pushed back on certain aspects of both. Without the proper guardrails in place, thinking machines could help companies or governments manipulate individual citizens, but the analytical and predictive capabilities of the cognitive machines that underpin predictive policing and social credit can also keep streets safer or establish financial trust in a country without an existing system of financial credit scores. For patients who want to participate, it could help nudge diabetics toward healthier lifestyles and reduce relapse among opioid addicts, as two of the ten finalists for the IBM Watson AI XPrize are setting out to do. It can help process the massively complex atmospheric problems and shore up existing climate change simulations to help us better understand the danger human activity poses to the environment and how to mitigate it. It can help reduce deforestation and illegal logging, as the Rainforest Connection does with solar-powered smartphones, called guardian devices, that it mounts in trees. The devices monitor the forests by sending sounds up through

---

* Interview with the authors at the University of Texas at Austin, October 20, 2017

the local mobile-phone network to its AI system in the cloud, which can identify the sound of trees being felled and alert a local response.*

It can even help eliminate so much of the tedium that clogs our everyday lives. After selling his last company in 2013, Dennis Mortensen went back through his calendar to see how many meetings he'd scheduled the prior year. All told, he'd arranged 1,019 meetings, and he had to reschedule more than 672 of them. "I'm 45 years old, in the workforce about 20 years," Mortenson says. "This view of the future where I'm doing another 20 years of sitting around in some version of an inbox playing email ping-pong, it didn't seem real to me." So, he founded x.ai, a startup that developed an AI scheduling assistant that can schedule meetings on your behalf. He used it to set up his interview for this book, and it worked seamlessly—a little notation on the email address noting the difference between a human response and an interaction with the bot. "The more I looked at this particular chore, the more I realized it shouldn't have been a human job to begin with, but we do plenty of those," he says. Finding ways to automate those tedious tasks and free our potential "is the only way we can move forward."

With quickly growing deployment of more sophisticated AI systems, we might gain a vast new control over the tedium of our lives, the threats to the environment, and the level of danger on our streets. Yet, those same systems can go too far and threaten our individual agency and lives. A new power balance will emerge in the coming era of thinking machines—between countries, companies, people, and machines. Already, the world's two leading AI superpowers, the United States and China, are providing a glimpse of how these different power balances might shape up in the years to come.

SOCIAL COHESION AND SOCIAL CREDIT

For years, John Fargis and his wife, Vida, would return to Shandong to visit her family and, as is customary in one of the most-traditional regions of the country, celebrate the Chinese New Year with veneration for those who came before them. Families welcomed their ancestors on

---

* Topher White, "The fight against illegal deforestation with TensorFlow," Google blog, March 21, 2018.

the first night. They ate dumplings with wheat grown from the soil of their predecessors' graves. A visit to a friend's or neighbor's home elicited an enthusiastic recitation of the host family's lineage. And on it went, until the third night popped with fireworks, ushering the ancestors back to their graves. To this day, the region's culture remains deeply rooted in Confucian ideals of filial piety and structured society. "It's deeply traditional, powerful, and rich," says Fargis, a professor at the Hult International Business School who's spent much of the past twenty-five years in China, including as Henry Luce Scholar and the first foreigner to gain permission to teach in a Chinese reform school setting.*

Confucianism, that "absolutely exquisite, rich and opaque tradition," spun the essential threads of Chinese culture for more than 2,500 years. It held up for millennia, against the influence of nearly every foreign philosophy of life or authority. Buddhist thought migrated in from India during the Han Dynasty, around the first century BCE, but it did little to change the central cultural principles of Confucianism. In fact, Buddhism quickly got "China-ized," Fargis says, and the conventional wisdom came to regard it as a philosophy seeded by Chinese philosopher Lao-Tzu, the creator of Taoism. Centuries later, around the time of the US Civil War, the Qing dynasty put down the Taiping Rebellion (1850–1864), which was sparked by a Christian millenarian sect whose leader, Hong Xiuquan, considered himself the younger brother of Jesus Christ. Throughout its existence, Chinese civilization did a more thorough job of either resisting foreign cultural impregnation, or culturally appropriating those external influences, than virtually any other civilization on the planet, with Confucianism as the bonding internal force.

Yet, by the early 1900s, a growing chorus of criticism began to question the Confucian notion of virtue as the central element of orderly society. The last Confucian exam took place in 1906, seen as a relic of a bygone era and a vestige of a system that couldn't quite explain a world order in which China was down with the world's foot on its throat. These imperial examinations, which had been in place for centuries, served as a gatekeeper to government jobs. However, they mainly tested knowledge of Chinese classics and literary style, doing more to maintain the country's shared cultural, intellectual, and political sensibilities than to ensure an appropriate level of technical and practical expertise in

---

*   Interview with the authors in Berkeley, CA, March 2, 2018

the state bureaucracy. The conclusion of these official examinations and the subsequent end of the Qing Dynasty in 1911 deepened the erosion of China's Confucian construct and opened the country to the embrace of a new import of foreign thought: Communism. But even then, the deeply rooted traditional ideals of a well-ordered society based on hierarchy, orthodoxy, obedience to authority, and discipline were fortified in a cryptic way by the introduction of Leninism, says Orville Schell, director of the Center on US-China Relations at the Asia Society.* Just as the imperial dynasty remained the organizing principle of a highly striated society during traditional times, the Party became the preeminent centralizing force during China's revolutionary Communist period and remains so today during its more recent emergence as a geopolitical and economic powerhouse. And in both the ancient and modern scheme of things, the notion of the importance of a "big leader" at the heart of China's political and societal environment and a strong unified one-party leadership has endured, Schell says.

More recently, President Xi Jinping's rise to power re-enforced these notions of the centrality of a single power center guided by an authoritarian "big leader," a role that Xi has more than fulfilled since his reign began in 2014, Schell explains. With major programs such as the Belt and Road Initiative (BRI), which envisions trillions of dollars of investment in seaports, roads, railways, and other major infrastructure project across Asia, Europe, and Africa, Xi has written himself even larger than Mao and Mao's later Party successors. Xi has reinserted more government control and influence in both political and private-sector life as China seeks to enlarge its global footprint and become a new kind of great power "with Chinese characteristics."

Yet, as Fargis notes, within the bounds of Party authority, Chinese citizens enjoy a certain freedom of activity and, for so many, a soaring standard of living that most couldn't have dreamed of just a couple decades ago. And that has started to change some of the most deeply held Chinese beliefs, even in traditional Shandong. When Fargis and his wife inquired about the usual holiday visit in 2017, his brother-in-law said he planned to take his family on a vacation to Hainan island in the South China Sea, far away from their home village, instead. "If you think about what the missionaries couldn't achieve with their

---

* Interview with the authors in Berkeley, CA, March 9, 2018

efforts to convert folks in Shandong to Christianity; what the Japanese military couldn't achieve; [and] what Maoism couldn't achieve to erode Confucian tradition," he says, "arguably, technology-driven prosperity is now achieving."

This comparatively new prosperity is chipping away at a deep cultural foundation, even as it is expanding upon an incomplete economic foundation—one that presents possible drawbacks but opens even more tremendous opportunities. For example, most urban Chinese use Alipay and similar mobile phone apps to pay for things, substituting the touch of a button for cash and the credit cards that most citizens never had in the first place. China already has moved well beyond most of the Western world when it comes to mobile payments. Viewed from the perspective of a nation rushing rapidly forward while adhering to the Party's vision of its tradition, culture, and economy, the country's emerging social credit system makes more sense.

CREDIT WHERE CREDIT IS DUE

Alibaba's Zhima Credit (or "Sesame Credit") emerged as one of several pilot programs for what's expected to eventually develop into a national social-credit program that would include more than just financial information. The real push behind the initiative began in January 2015, when the People's Bank of China issued eight provisional licenses to private companies, hoping to encourage more widespread access to and use of credit scores. While the companies said they'd made progress, the PBOC had not fully licensed any of the eight systems three years later, and more than half of China's citizens still did not have an extensive enough personal financial history to borrow from formal financial institutions, according to a policy brief from the Peterson Institute for International Economics.* That left in place a vast, informal lending network throughout the country, and Beijing had little insight into how indebted its citizens were and whether repayment trends were improving or worsening. So, absent the benefit of an established credit rating system like the one in the United States, they sought to create

---

* Martin Chorzempa. "China Needs Better Credit Data to Help Consumers," Peterson Institute for International Economics policy brief (January 2018).

one by scraping e-commerce, social media, and other online data from applications such as WeChat and Alipay.

Yet, the service aggressively expanded beyond mere financial credit "to serve as a stand-in for an individual's trustworthiness," the Peterson Institute brief noted. Zhima Credit gathered a wide array of financial and personal data to generate a score on an 800-point scale for individuals. In addition to information about someone's online shopping habits and ability to pay bills on time, it collects demographic data and associations gleaned from social networks. So, a twenty-eight-year-old pregnant woman might receive a better "rating" than an eighteen-year-old man who buys a motorbike. Someone who has 700 Sesame Credits would be considered extremely respectable, whereas a score of 300 could trigger a range of social repercussions—anything from restrictions on first-class train and international air travel, to acquaintances ostracizing them to avoid being tainted by association. It eventually became one of the most extensive individual data-collection efforts the world had ever seen, with almost 200 million Chinese residents enrolling and being scored on their trustworthiness in transactions and relationships to date.[*] And that's before the system even becomes mandatory in 2020.

These and other smaller social credit programs pulled in a remarkable array of data types as well. Even before any purchasing or lending activity takes place, factors such as age, parenting, and social network would yield differences. One local program in Zhejiang even encouraged people to report their neighbors' breaches of social norms. The program, called "Safe Zhejiang," awarded discounts at high-end coffee shops or other perks in exchange for reports on anything from traffic violations to domestic disputes. Launched in August 2016, it reportedly had about 5 million users by the end of the following year, but it had met with stiff resistance from most residents who didn't care to be forced into surveillance of their neighbors or feared retribution for reporting concerns. Even some of the authorities in the province balked, worried it might undermine them.[†]

---

[*]   Mara Hvistendahl, "Inside China's Vast New Experiment in Social Ranking," *Wired* (Dec. 14, 2017).

[†]   Jeremy Page and Eva Dou, "In Sign of Resistance, Chinese Balk at Using Apps to Snitch on Neighbors," *Wall Street Journal* (Dec. 29, 2017).

Despite occasional resistance to some of the early initiatives, the push to develop the national social credit system continued unabated into 2018. Mainstream Western media portrayed the system as a means to monitor and control the population, especially minorities like the Uyghurs, a Turkic people of mostly Muslim faith in Xinjiang, a semi-autonomous province in China's outer northwestern region. Reports have shown that conspicuous activities like congregating groups leads to deductions in the point system. Local law-enforcement officials can access an individual's status through smart eyeglasses that identify citizens through facial recognition and, thus, facilitate more arrests.[*]

Yet Western concerns about the invasiveness of these technologies and the government's use of them to nudge people's behaviors don't raise widespread concerns among the Chinese populace. For one, the social contract underlying this use of smarter technologies is different in China, where the long-standing Confucian tradition of deference to authority in the name of stability still holds sway. And as multiple Chinese academics, AI developers, and entrepreneurs noted in interviews, today's residents are more optimistic about the power and potential beneficial aspects of advanced technologies. "If you look at the Chinese history, just in the last 40 years, . . . the people who embraced change benefitted the most," said Ya-Qin Zhang, president of Baidu and the former head of Microsoft Research Asia. "Plus, there's a constancy of direction from government, so even people who come back to the [technological] change later are winners." And because of that embrace of technology—along with a general sense that the most troubling aspects of AI remain distant possibilities—little discussion of the potential downsides occurs in China's mainstream media or public discourse, although fluent global information flows will soon bring more of those issues to the forefront, says Hsiao-Wuen Hon, the head of Microsoft Research Asia.

Yet, even for those who accept the concept of the social credit score, unresolved questions remain. For one, experts suspect a social credit system could lead to the formation of new social classes, as people with good ratings shy away from those with poor ratings. More immediately, though, what sort of recourse do citizens have to rectify mistakes or otherwise challenge their scores? Even in the United States, where credit

---

[*] Josh Chin, "Chinese Police Add Facial-Recognition Glasses to Surveillance Arsenal," *Wall Street Journal*, Feb. 7, 2018.

agencies Experian, TransUnion, and Equifax hold oligopoly positions, recourse to address poor or erroneous credit scores is spotty at best. In fact, simply requesting a copy of one's credit report more than three times in one year triggers a score reduction. And, contrary to the complex procedures and opaque rating systems employed by the US systems, the Chinese government has openly stated its approach and philosophy: Counteract corruption and other untrustworthy patterns of behavior while fostering greater reliability in economic and interpersonal transactions.

But, viewed through the lens of western ideals of civil rights, as the AI-powered social credit system closely examines, quantifies, and makes public not just criminal or financial activity, but demographic characteristics and the "rough edges" of interpersonal relationships, how close does it get to becoming a digital branding iron? Participation currently is voluntary, but the government already has put restrictions on travel for those who fail to pay bills or engage in other unsavory activity. If it becomes mandatory in 2020, as initially planned, how will it affect the economic existence and social dynamics of almost 1.5 billion Chinese citizens? At some point, too much digital control might trigger unintended consequences that decrease, rather than increase, stability.

PREDICTIVE POLICING

In late 2017, hoping to illustrate just how vast a surveillance network Chinese authorities have built, BBC reporter John Sudworth requested and was granted rare access to one of China's high-tech law-enforcement control rooms. Officials agreed to temporarily mark Sudworth as a person of interest in their database and then, loose on the streets of Guiyang without being physically shadowed, he set out to see how long he could avoid detection and apprehension. In a city of more than 4.3 million residents, he lasted just seven minutes until police nabbed him.*

Dozens of countries around the world have started deploying a range of AI-powered technologies to monitor people, everything from facial recognition to gait analysis to speech-pattern analysis. While most Chinese citizens expect and even shrug off widespread surveillance initiatives

---

* Joyce Liu (producer), "In Your Face: China's all-seeing state," BBC News, Dec. 10, 2017.

there, few US residents realize just how personal and how deeply integrated AI-backed surveillance has become in many of the country's largest cities. In some parts of Los Angeles, for example, street cameras capture the faces of people in the vicinity when crimes were committed, logging innocent passersby into the system. Cross paths with criminal activity more than once, and the score behind a person's name in the system goes up, heightening their potential interest to police. This kind of correlation is helpful for police, who don't allege that presence means involvement in illegal activity. But it can lead to troubling situations. In one case in Fresno, California, an AI system used by authorities assigned threat levels to residents based on social media posts and billions of other commercial records. Officers responding to incidents would see conclusions drawn by algorithms that public citizens couldn't review, according to Matt Cagle, an attorney at the American Civil Liberties Union (ACLU), which is suing the LAPD. Speaking before a California Assembly committee hearing, Cagle said: "When one City Council member found himself flagged as an elevated threat, he had no way to determine the basis for that decision. There were no rules governing the use of that system." The lack of clear guidelines allows officers to label or stigmatize residents, he and the ACLU argue, and no regulations exist to require the LAPD and other forces to expunge records, anonymize faces, or label individuals as mere contextual elements in a crime area, like buildings or trees instead of potential suspects.*

When she set out to study the LAPD's use of its intelligence-driven policing, Brayne, the University of Texas sociologist, saw ample evidence of retroactive data surveillance. It naturally occurred as investigators went to the various databases to research potential suspects and construct an argument for search and arrest warrants. But in some cases, she says, that information never appeared in affidavits or in evidence. It was, she says, "rendered invisible when it [was] submitted to the courts." She would see it happen in the field, but then not see evidence of it in the courtroom.†

Yet, what surprised her even more was the skepticism many law enforcement officers expressed about the systems, which could

---

* Testimony to California Assembly joint hearing of the Privacy and Consumer Protection Committee and Select Committee on Emerging Technologies and Innovation, March 6, 2018.

† Interview with the authors at the University of Texas at Austin, October 20, 2017

integrate a range of officer-monitoring technologies, such as GPS tracking of their patrol cars. As a scholar of the criminal justice system, Brayne expected officers to embrace the "information is power" aspect of data intelligence, but many saw it as an entrenchment of managerial control. In fact, the Los Angeles Police Protective League, the union of LAPD officers, has resisted the use of a range of monitoring technologies, many of which, while available, aren't used.

Few Los Angeles and New York City residents can wield the same measure of resistance against these technologies, in part because neither the LAPD's nor the NYPD's predictive policing systems were proactively communicated to the communities, much less debated in a wider public forum.[*] That shouldn't come as a complete surprise, as police departments playing catchup with tech savvy criminals don't want to tip off criminals about their enhanced capabilities. As Jonathan Feldman of the California Police Chiefs Association put it in a public hearing in front of the California State Assembly: The broad public, including criminals, already have all these technologies, so how are police supposed to keep citizens safe if they can't use the same or better tools to connect the dots on illicit activity? And if we debate everything the police do publicly, we tell the bad guys how to avoid detection.[†]

Yet, there's an inherent open-or-closed tension within law enforcement itself. Law enforcement officials also want the public to know about mass surveillance, so it can act as a deterrent to criminal activity in general, Brayne notes. "Part of the thing is to communicate to these guys on the street, 'Hey we're following you. We know who you are and who your affiliates are and where you hang out and what you've been up to. So, don't do something illegal because we are already on to you,'" she says. "If you don't ever need to intervene that's the most effective law-enforcement mechanism."

That's especially critical in an age when ever "smarter" digital technologies offer reach, skill, and anonymity all at once. Anybody with a mobile phone can access advanced digital tools and inflict reputational and material harm on others with very little accountability. In this

---

[*] Barbara Ross, "NYPD blasted over refusal to disclose 'predictive' data," *New York Daily News*, Dec. 17, 2016.

[†] Testimony to California Assembly joint hearing of the Privacy and Consumer Protection Committee and Select Committee on Emerging Technologies and Innovation, March 6, 2018.

environment, law enforcement needs the capability to not just track, but ideally prevent perpetrators from acting. But they're damned if they do and damned if they don't. Americans might despise the idea of unfettered surveillance, but the fact that federal agencies had pertinent evidence yet couldn't connect the dots, share information across departments, and prevent the 9/11 terrorist attacks might bother many of them even more. The 9/11 Commission Report released almost three years later found that various US authorities collectively had the information they needed to identify and stop the terrorists but had no systems in place to share that information. It recommended the creation of a national intelligence director to help coordinate intelligence between agencies.

Surely, tourists feel safer traveling through New York City today, now that authorities can better track information, predict potential incidents, and, hopefully, defuse tensions and prevent criminal activity and harm. If their primary purpose is to prevent as much injury and death as possible, dispatching first responders after a crime occurs amounts to a failure of duty and a significant drain on human and economic resources. And that doesn't even begin to touch on the subsequent human and monetary costs of trials, probationary proceedings, and incarceration. Judges in parts of the United States already use AI systems to help determine eligibility for bail and similar probationary questions. Yet, neither the systems nor their developers can (or will) explain the machine's reasoning, often citing the need to retain trade secrets in a competitive marketplace. This is a problem in countries purportedly based on the transparent rule of law. If we can't explain the reasoning, then defending attorneys for an alleged criminal can't argue against it, diminishing their ability to provide an alternative narrative for the defendant.

But beyond questions of law, we might ask deeper questions about how this influences the formation of human judgment. Judges and juries bring their intellectual and analytical skills to bear in the courtroom, but also their notions of social equity and empathy for both victims and offenders. The "jury of peers" is a fundamental strength of the US judicial system, in part because people can contextualize the offending act. At present, developers can't bake such a full array of human contextualization into machine learning algorithms, which are most often designed by programmers thousands of miles away from a given situation. So, even as the US judicial system seeks ways to remove tedium and bias from some of the tasks required of the bench, how will it digitize our empathy and a sense of our shared, often-conflicted social values?

SERVICE DELIVERY, OR PUNITIVE ACTION?

It's the type of question worth asking, because the upsides and downsides both could transform society. Predictive policing could establish more civility in communities, fostering an atmosphere that's conducive to business, tourism, customer traffic, and everyday life. By protecting our property and personal well-being, it could help attract greater investments in neighborhoods and social goods, building the types of wealth that secure families and help fund schools and other community amenities. We might even refer to those outcomes of predictive policing as "trickle-up" urban development. But we also need to think about the potential for that same trickle-up development to generate rampant gentrification and the marginalization of people who are "digitally less hygienic" or otherwise disadvantaged because of their socioeconomic status. We need to consider how to integrate AI and other advanced technologies with broader urban- and social-development policies, so these innovations don't amplify existing problems.

As Brayne observed, the LAPD's predictive policing initiative already prompts different reactions to individuals, even for the same types of incidents. Police typically adopted a reactive mindset when responding to calls for domestic violence in neighborhoods with a lot of gang, drug, or other criminal activity. Yet, a similar call from someone with a house in an affluent neighborhood and other positive attributes culled from background databases put officers in a "service delivery" mindset. "In rich communities, a lot of the calls for service would be people threatening to kill themselves," she says. "In those cases, they would look to see what might be missing here. Do they already have linkages with child and family services? Did they recently get divorced? Did they lose their job? It's very 'service delivery,' rather than 'incriminating.' . . . In an area with a lot of gang activity and crime, they look to see if this woman is also involved in a criminal justice system, like maybe she's out on parole. It is service delivery oriented when there are kids involved, for sure, rather than just punitive. But yes, I don't know if I saw a lot of examples of it disconfirming biases."

While the ACLU and some LA-based organizations have started to push back on AI-based predictive policing, discussions about how to govern these systems and what values the community wants to instill in these systems remain, at best, preliminary. Broader public awareness isn't much better. In fact, Chinese citizens might be more broadly aware of and in tune with the social credit system than Americans are with predictive policing activity. So, where will this lead in both China and the United States? What might build trust in one country could erode it in the other. Citizens might start to avoid areas with camera surveillance, shunning targeted neighborhoods and depriving them of commercial or civic activity. The same systems design to reduce crime might increase bias and social stigma and create a database that supports the dismantling of stereotypes about minorities.

The fact is, predictive analytics, AI, and interactive robotics already have become essential and valuable tools for individuals, governments, and businesses. To enhance our lives and our communities, however, we need to operate these technologies within a shared set of community values, to balance the power of their predictions with the people affected by them, and to establish a new threshold of trust, lest we forget that trust is *the* most valuable currency in a society. But unfortunately, if unsurprisingly, the public and political debate about AI's influence on a variety of societal power balances will advance far more slowly than technology does.

BACK TO THE FUTURE SHOCK

While pockets of expertise have sprung up throughout the world, the United States and China have emerged as the undisputed leaders of AI research, development, and deployment. Yet, even these two countries—perhaps *especially* these two countries—have seen the pace of technological development advance beyond the regulatory, legal, or ethical frameworks needed to govern the role of thinking machines. Humanity has never stopped a sector of technological development cold. While different countries have adopted moratoriums or bans at various times—on cloning or chemical and nuclear weapons—we continue to advance the state of the art in genetic engineering, weapons, and virtually every other technological category. Humans consistently advance. We experiment, fail, and hurt. And then we shape, optimize, and perfect technologies, only occasionally or more slowly mitigating their risks with multinational agreements and monitoring institutions.

Crops of genetically modified organisms (GMOs), for example, provide significant benefits for broad swaths of the global population. Modified plants engineered in the lab can provide greater pest and drought resilience, lower cost and more affordable production, and enhanced control of water and other resources, especially in impoverished regions of the globe that are fighting starvation. Secondary economic benefits accrue, as well—from industrial-scale job creation in the development, rearing, and harvesting of crops, to the shareholder returns for retirement funds invested in GMO companies. But many people suspect that GMOs might not be healthy for us or for the natural ecosystems of crops. European citizens feel more strongly about the danger of such organisms, in part because of a general distrust of humanity's manipulation of nature and of large corporate interests, while people in less-developed nations might see the direct advantages of GMOs as they struggle to support themselves with less-robust natural crops.

We see similar arguments play out in anything from arms control to the global production and consumption of fossil fuels, with local historical experiences, divergent philosophies about economic growth, and varying degrees of consensus about viable alternatives driving different debates in different places. Europe, for example, suffered through two catastrophic wars and various dictatorships, some of which conducted experiments on human genetics that engrained a deep belief that interference with nature is morally reprehensible. A more laissez-faire attitude toward corporate innovation and capitalistic pursuits, including GMOs, makes sense in the United States, which never suffered a similar direct trauma. We all innately realize any technology can carry dual purposes—a hammer can pound a nail or crush a skull—but our human imagination can conjure vastly different images of the potential harm to ourselves and our societies.

Even when we fail, dampening our enthusiasm for growth and giving us a proverbial bloody nose, we keep pushing forward. The dot-com implosion hardly even slowed the web's increasingly perva-sive, influential, and powerful role in our lives. This inexorable drive forward raises critical concerns, particularly in relation to artificial intelligence, where one has no problem finding doomsayers or imag-ining grim futures. This sort of dystopia has longstanding precedent in science fiction and literature, from *Frankenstein* to the industry of futurism ignited by *Future Shock,* the 1970 phenomenon and best-seller from Alvin Toffler, who argued that such rapid change in too short a time would overwhelm people and societies. Futurists and sociologists have always warned us and prescribed ways to prepare for the dangerous effects of accelerating change. The same prophets talking today about the potential for AI to create a feudal structure of workers—split between owners and the gig workers trying to latch on with them—echo Toffler's concerns about a shredding of the economic fabric and the creation of a temporary-worker underclass.

These visionaries might very well be right, but such concerns about the dangers rarely translate into proactive decisions to take back control over development and check the power of technological development over our lives. Apothecaries made medicines and poi-sons, eventually producing powerful opiates long before recognizing the potential dangers of abuse. When innovations produce what we regard as harm to individual people, we do little to act. Only when they became major health epidemics—cigarettes causing widespread cancers or opioid addictions gripping more affluent or privileged sec-tors of US society—did policy makers begin to call for action. One of the only examples of human foresight and proactive guidance of tech-nological development applied to the genetic engineering of humans, where the UN member countries adopted guidelines condemning it before it was openly exercised.

AI IS FASTER AND MORE POWERFUL

In many circles, efforts to guide the development of artificial intel-ligence generate a similar sense of urgency. For one, its pervasive influence on our lives, values, and relationships could transform human societies, cultures, and economies far more radically than even the Internet has. Then, much like the Internet, AI-based innovation can occur at incredible speeds, even if it comes in fits and starts. It has taken seventy years and a few chilly "AI winters" to even get to this nascent point where we stand today, but a few breakthroughs in deep-learning techniques, the explosion of large data sets, and the availability of cheaper computing power unleashed a breakneck pace of new AI applications, often surprising experts within the field itself. No one expected AlphaGo could beat Go world champion Lee Sedol in a five-game match in 2016, yet the neural network won four of the games. Then, about eighteen months later, AlphaGo Zero came along and, having taught itself the game, beat its predecessor in a hundred consecutive matches.

Each instance of these rapid changes that catch us off guard raises difficult questions that need answers not just from technologists, politicians, or regulators, but from society as a whole. As Burkhard Huhnke, vice president of automotive strategy at Synopsys, a Silicon Valley software and microchip developer, and former senior vice president of e-mobility and innovation at VW America, explains, our roadways become far safer if we remove error-prone humans from behind the wheel. But the idea of telling people they can't drive anymore is a different issue altogether, says Huhnke. That's "a more social aspect," he says. "Because it touches the liberty of being a driver yourself. . . . This can't be fixed by regulators; this has to be figured out in a completely different dimension."*

GETTING UNDER OUR SKIN

The unusually close connection between the PARO robotic seal and its users might begin with its sheer cuteness, but it's the subtle design touches that really secure the bond. Under its soft white fur, developers placed the robot's touch sensors in balloons so users wouldn't feel hard spots. Its big black eyes follow movements around the room. It moves its flippers and gurgles with attention. It changes its body posture in response to human touch and it knows the difference between being stroked or rough-handled. Even when recharging, its power cord makes it look like it's sucking on a pacifier. That adorability belies

---

*   Interview with the authors via video conference, January 18, 2018 the serious research and development that made PARO such an effective aid for caregivers serving elderly patients, especially those with dementia and Alzheimer's. Developed by Takanori Shibata, a chief research scientist at Japan's National Institute of Advanced Industrial Science and Technology (AIST), the artificial intelligence underlying PARO adjusts its behavior to its interactions and surroundings. It responds to its name and its users' most common words and, if it goes long enough without affection, it starts to squeal.

That combination has imbued PARO with a remarkable ability to help caregivers soothe disoriented elderly patients and help them communicate. After a 2008 study determining its effectiveness, the Danish Technology Institute encouraged every Danish nursing home to buy one.* Since its introduction in 2004, thousands of the PARO seals were put into use in nursing homes in Japan, Europe, the United States, and elsewhere. In the weeks after an earthquake and massive tsunami wrecked Japan in 2011, a pair of donated PARO seals helped sooth nursing home residents in Fukushima, one of the country's worst-hit areas.†

Yet, for all the attention to the soft features and hard science that went into PARO, the depth of its effectiveness stems from something far more human: our ability to suspend our disbelief. Shibata and his team made PARO feel close to real without falling into the "uncanny valley"—that odd place where a lifelike robot begins to feel a little too creepy. So, while they made PARO to weigh about six pounds, around the weight of a baby, they also deliberately made it a seal to avoid existing associations people have with common pets like dogs and cats.‡ We know the seals aren't real, but they're just real enough in our heads to make us happy and interact on a deeper level with them.

Jonathan Gratch, a professor of computer science and psychology at the University of Southern California, spends much of his time researching the extent to which people treat machines as social beings. In other words, he thinks deeply about what one might have to add to a robotic seal to make people relate to it more intensely and treat

* Wolfgang Heller, "Service robots boost Danish welfare," Robohub.org, Nov. 3, 2012.

† https://www.youtube.com/watch?v=PNw4oicWmWU

‡ Andrew Griffiths, "How Paro the robot seal is being used to help UK dementia patients," *The Guardian* (July 8, 2014).

it more like a real seal. What he's found is that people initially tend to treat interactive robots or other AI systems like social entities, granting them a certain measure of empathy. And when the machines provide cues that suggest an emotional connection—the more it appears to emote, for example—they can provoke even stronger response in their users. Developers of AI and robotic systems often program them to convey these humanlike attributes or emotions, which can foster a tighter bond with the machine. However, over time, those emotional cues need to have meaning behind them, and many of these designs eventually run into what Gratch calls "false affordances."

For example, consider a robot or chatbot that appears more human-like because it can direct one's attention to a specific object or an idea. Research shows that sort of attention deepens interactions with humans. But when the system tries to direct a person's attention to an irrelevant or confusing target, people quickly lose faith in it, Gratch explains. The same holds true for attempts at empathy and shallow apologies. So, while people typically over-trust the machine at the outset, they perceive it as an outright betrayal when it doesn't fulfill its function or satisfy its promise. "If a machine does recognize that and improve," Gratch says, "it's very powerful because it's following 'the rules.'" In essence, it sustains, and perhaps even heightens, our suspension of disbelief.

This prompts some concerns among ethicists—initial reactions to PARO included worries that it misled and manipulated patients, for example—but we have any number of harmless examples in our lives. We tell our kids about Santa Claus and encourage them to make believe with their teddy bears and toys, and we revel in the creativity they express when doing so. We take deep pleasure from movies, theater, and novels that inherently require our own imagination and a grain of salt. And we readily employ it in our interactions with technology, including in ways that accelerate the development cycle. We're willing to accept the fact that our smartphones have "planned obsolescence," meaning that they are never going to be "done" or "perfect," with the next itera-tion coming along soon enough. The changes that occur, whether to our benefit or detriment, often are subtle and go unnoticed, but they are just enough to please us and bridge the time to "the next big thing."

This accelerates and amplifies development of machine cognition even more, because these AI systems regularly address some of our most fundamental, immediate, tangible, and common needs. This makes it far easier to "suspend our disbelief" about imperfections, so to speak, in exchange for the benefits we reap. We're willing to turn over more and more of our cognitive workload to something far from perfect. Eventually, when we step back and look, the pervasiveness and degree of disruption and possibility in our lives will surprise us. We have allowed the machine to get under our skin and into our heads, and from there it can exert more subtle power over our lives. That could provide great benefit, helping alleviate stress among elderly citizens, but we risk conceding more control than we intend.

BENEFITS AND SIDE EFFECTS

The promise and threat of AI applications feel more urgent because of the unknowns inherent in such a rapidly developing technology, but we humans have a remarkable ability to accept tradeoffs that don't always favor our best interests. Nuclear power in the 1970s provided a cheap and clean source of electricity. The victor's narrative of shortening World War II had cleansed US nuclear power of some destructive moral baggage, and officials heralded it as an alternative to energy shortages and fossil fuels. Iconic scientists like Albert Einstein and Niels Bohr helped establish nuclear power's reputation as the energy source of the future, and it began to take a place in the lineup of other innovative technologies that ruled the day: the automobile, the television, and the telephone. A shiny future had arrived. We saw the promise and we suspended our disbelief about the awesome power of atomic technology.

For Americans and many others around the world, that flirtation ended abruptly on March 28, 1979, when Unit Two at Three Mile Island (TMI) in Pennsylvania partially melted down. Nuclear power's negative side effects became immediately and starkly clear: It could kill large amounts of people in one major accident. The incident in Pennsylvania brought back the fears that attended Hiroshima and Nagasaki, anxieties that have been heightened in the decades since with accidents at Chernobyl and Fukushima. Today, there is nothing subtle about nuclear weapons or nuclear power.

While generally considered far less controversial and catastrophic, the automobile has gone through a similar cycle. Its substantial social and economic effects were visible from the outset. It brought freedom to individuals and families, and it remade community and economic development—sparking suburban growth as people became more mobile, while also creating a vast automaker ecosystem with tiers of suppliers, distribution channels, and ancillary services. Today, most developed nations consider auto industries too large, too important, and too strategic to fail.

When we step back and consider the evidence, though, the automobile has caused greater harm to the environment and our lives over the past century than nuclear power. Of the 587 million metric tons of carbon dioxide and equivalents emitted by the United States in 2015, more than a quarter came from transportation, second only behind electricity production, according to the Environmental Protection Agency (EPA).* With the noise, stress, and lost productivity, the ancillary costs of the car already begin to add up and its status as a symbol of independence and freedom wanes. Of course, the direct human cost is even more dire. Cars and trucks and their fallible human drivers kill almost 1.3 million people annually around the world—the ninth-leading cause of death, sandwiched between diarrheal diseases and tuberculosis—with an additional 20 million to 50 million people injured or disabled, according to the Association for Safe International Road Travel.† In the United States alone, more than 35,000 are killed each year, creating an additional financial cost of $230.6 billion a year, the association says.

There is nothing subtle about the concurrent promise and threat that accompany automobiles and nuclear power. And the basic premise of both have remained essentially unchanged for decades. In early 2018, about 450 nuclear power plants provided about 11 percent of the world's total electricity generation, according to the World Nuclear Association,‡ and an estimated 94.5 million new light vehicles were sold worldwide in 2017, according to research firm IHS Markit.§ And

---

* U.S. Environmental Protection Agency, *Sources of Greenhouse Gas Emissions*, Updated April 2018.

† Association for Safe International Road Travel, *Annual Global Road Crash Statistics*, ASIRT.org.

‡ World Nuclear Association, *World Nuclear Performance Report 2017*, Updated April 2018.

§ IHS Markit, "Global Auto Sales Growth to Slow in 2018, Yet Remain at Record Levels; 95.9 Million Light Vehicles Forecast to Be Sold in 2018, IHS Markit Says," press release, Jan. 11, 2018.

we pump hundreds of millions of dollars into research to make both technologies safe and ensure that most of us never directly experience that potential for physical or psychological damage.

Unlike nukes and cars, artificial intelligence and its cousin technologies don't display obvious, visible effects on our everyday lives. Even when they do, the side effects of those technologies in our lives are far less clear—and in that sense are more akin to pharmaceuticals. We saved hundreds of millions of people from certain death by drastically reducing the incidence of polio and whooping cough and arresting the fatality rates of HIV/AIDS. We fed millions more with high-protein, high-carbohydrate engineered diets so they could grow stronger, live longer, or enjoy enhanced well-being. Seen from the perspective of an average day in our lives, every decade since World War II has brought advances in medicine and agriculture that have made those days richer and better. Scale economies of industrial agriculture brought greater affordability in food and more conveniences in preparing it. Restaurants and equally scaled industrial-style restaurant chains have made food more accessible, affordable and, in many cases, more pleasurable.

Yet, these advances have sent subtler and more-pervasive ripples through our health and our societies, as well. As the use of antibiotics in humans and livestock exploded, more types of bacterial infections mutated and adapted to the drugs, sparking an arms race between pharmaceutical companies and Mother Nature. Every time we go to the next round of drugs we risk launching successive waves of infection by new mutations of anything from flu strains to Strep bacteria. Medications also spun off a range of addiction crises as opioids, antidepressants, steroids, and other drugs became more commonplace in our treatment plans. As we tried to optimize our health, a multibillion-dollar supplement industry emerged despite the absence of scientific proof of efficacy. In so many cases, we don't truly understand what all this added biochemistry does to our bodies, so the consequences remain hidden and we largely ignore them in our daily choices.

The widespread use of artificial intelligence will enhance our humanity, our well-being, and our lives in so many ways, but we need to consider its potential side effects in the same way we think about pharmaceuticals, not the histories of automotive and nuclear power. The latter two technologies have had very visible and tangible effects, good and bad. By contrast, we can't see nor understand the often-complex algorithms and neural networks any more than we understand the tiniest and deepest machinations of biological bacteria and viruses and chemical interactions in our medications and bodies.

Those unseen elements can deliver invaluable progress and prosperity. A 2014 study by the Centers for Disease Control and Prevention (CDC) estimated that vaccines prevented more than 21 million hospitalizations and save 732,000 lives among the children born in the previous twenty years, and that was just for the United States alone. With data as its fuel, AI is poised to become the engine of a new and fruitful autonomy economy. But to harness that potential—to use AI to cure the ills of modern society and avoid the worst of the possible side effects—we need to bend the power of these systems toward the benefit of humanity. And to build the necessary guardrails, we need to understand the subtle, unobservable, and tremendously influential control these thinking machines will exert on us as they seep into our lives.

SEVEN DEGREES OF POWER

Shaoping Ma reads some of the same artificial intelligence hype and prophecy in China as his peers do in the United States and Europe. Usually, the Tsinghua University computer science professor says, it's the media trumpeting concerns of self-driving cars endangering pedestrians or malfunctioning robots injuring human workers. The mania rarely seeps very deeply into the broader consciousness, Ma says through an interpreter, but it spreads through WeChat and other social media apps. "The media wants exciting news," he says, "and that's true in China as well." The human mind can make tremendous imaginative leaps when it comes to technology, going further than even high-tech roadmaps. So, while there's little evidence of an emerging general superintelligence to date, Ma says, many Chinese residents worry about AI's development because they know so little about the real technology itself. Like anywhere else, most people in China don't realize how deeply AI-powered technologies have already permeated their lives.*

---

* Interview with the authors via video conference, December 12, 2017

Ma sees it at the university, in his role as vice chairman of the Chinese Association for Artificial Intelligence, and in his joint work with Sogou.com, the country's second-largest search engine, trailing only Baidu. Even with the government's announcement of a massive commitment to AI research and development in the coming years, the intricate ways thinking machines infiltrate and influence Chinese lives will remain subtle. In his work with Sogou.com, for example, Ma and his colleagues study search engine challenges, including the idea that people often want to discover something a bit different from what their actual search terms suggest. Like most top search engines, they developed their algorithms with "result diversity" as one of the major metrics, using it to satisfy different information needs according to different users' search intentions.

Your search for "wings" might ordinarily return a list of bird-related items that interest you, but it also might include references to the Detroit hockey team and Paul McCartney's rock band. Although your interest as a birder is known to the system, this variety of results assures a more consistent experience across users, reducing the effect of pure clustering. Result diversity and other subtle adaptations enabled by AI seep into most of what we do online with our computers and smartphones today. In fact, some core AI models, including those that categorize similar shoppers, film buffs, and news readers to improve recommendations and boost sales, have drifted toward commodity status, a basic building block for virtually every digital interaction we'll have in the years to come. The potential for applications range from medicine to construction, and will spread far wider as other more-analog domains grow increasingly digital.

For venture capitalists like Aaron Jacobson, a partner at NEA in Silicon Valley, the products researchers develop on top of commoditized AI platforms had become the more attractive investments by 2017. That follows the broader move toward a new "platform economy" built around and dominated by corporations, most based in the United States and China. These AI titans—Alibaba, Amazon, Apple, Baidu, Facebook, and Tencent—have harnessed the power of massive amounts of user profiles, transactions, and communications. They roll over every analog "brick and mortar" space, extending their reach and power to everything from our office layouts to our home thermostats. The sheer power of algorithms reveals itself in the sheer volume of people who use and subscribe to these global platforms. Facebook has more than 2 billion users. According to published reports, Amazon shipped more than 5 billion packages items through its Prime memberships alone in 2017. The scale in China is even more astounding, especially considering that these companies have started to approach or exceed similar numbers despite less global expansion than their US counterparts.

The power of these companies lies in the fact that they get to know their customers very well and become "sticky" in their lives. Once a platform's targeting of user needs reaches the right number of connections and the right strokes of satisfaction, people rarely delete their accounts and move elsewhere. Despite all the criticism that Facebook and Twitter took following the 2016 presidential race in the United States, they had little trouble sustaining their domestic user base. The fact that only one or two companies dominate social media, search, or e-commerce channels only helps lock in users all the more. This has led to quasi-monopoly status of some companies, whose subtle and powerful AI-driven platforms deliver more of what we want to heighten our reliance on their content and recommendations.

In fact, the stickiness of these Internet giants' algorithms is exactly what makes them so valuable across all their service offerings. AI-powered analyses of someone's search history sheds light on their interest in the next item to buy (hello, advertisers!), or the next person with whom to share their day. This kind of social-fabric weaving and reweaving make these titans such potent and helpful agents in the design of our lives. But the already commanding power they wield versus the rest of the economy and society keeps growing with every click and every new bit and byte that comes in. These companies can subtly and subconsciously nudge our decisions in invisible, irresistible, and often more profound ways than we care to recognize. Theoretically, we remain free to change our minds and our values—and we often do as our circumstances and environments change—but machine learning algorithms can play into our existing biases and reaffirm them without our conscious realization. We are subject to skillful manipulation that the average, non-technical person might never understand. Political operatives and skilled hackers can place fake news and advertisements into these social networks and subtly manipulate our perceptions of current news and emerging trends, as

Americans on Facebook and Twitter discovered in the 2016 presidential election cycle. AI will then help them place their well-cloaked content directly into the streams of unsuspecting users, heightening perceptions of certain news and exacerbating social tensions—even political outcomes.

Cognition is power, and the development of increasingly capable thinking machines represents a new source of power, a potent new intelligence that humans have created and with which they must contend. But in the end, we remain thinking beings, too. In the end, we interpret the media we consume, and we still cast our own ballots.

COGNITIVE POWER REMIXED

Unanimous AI launched in 2014 with the bold idea of creating "swarm intelligence," organizing a group of individuals into a sort of "real-time AI system" that combines and enhances individual intelligence and optimizes collective knowledge. The company's underlying algorithms help coordinate the diverse thinking in groups of people, combining their knowledge, insight, and intuition into "a single emergent intelligence," according to its website. Founder and CEO Louis Rosenberg likens it to the social interactions of bees, which resolve complex questions of hive location and food sources despite only having 100 million neurons in their individual heads. But the far more complex machinery of the human brain will reach its capacity limits eventually, Rosenberg said at the South by Southwest Festival (SXSW) in March 2018. So, why not continue to build intelligence by forming connections with other minds?

The process has produced better predictions and insights on everything from the Kentucky Derby to the Oscars. In fact, in early 2018, Rosenberg and his colleagues set the idea on the Oscars, creating an intelligent swarm of fifty everyday film fans and asking them to predict the winners. Individually, they were correct about 60 percent of the time. Collectively, they hit on 94 percent of their predictions, better than any individual expert who published predictions. Even more impressive, none of the group had seen all the nominated movies, and most of them had only seen about two of the films, Rosenberg told the SXSW panel attendees.

How many conversations played out across the country about the best movies to see in 2017? A friend recommends a movie, or a restaurant, or a product. That recommendation has value because it's supported by all you know or think you know about your friend—their standing and credibility and the quality of your relationship with him or her. Consciously or subconsciously, you assign a weight to that friend's recommendation. Your mind quickly examines your friend's trustworthiness, her experience with the subject at hand, and how many good recommendations she's made in the past. You know of the clear biases your friend might have toward or against the type of movie or the style of restaurant. Her profile and history are, for the most part, transparent to you. Ask the same of a few random colleagues at work, and your insight into their biases declines, as does their credibility in your mind. Get to people six or seven degrees away, and they have almost no credibility whatsoever—unless, of course, someone explains why you can trust them.

In theory, social networks could empower this sort of personal credibility, as their recommendation power emerges from the analyses of millions or billions of other people, from whom sensibilities like yours can emerge. At first blush, that sounds exactly like Unanimous AI, bringing together the power of multiple minds to help inform your own. But an algorithm that pulls insights out of collective data and fits the results to individuals merely represents a probability based on correlations with many other people, and not an individual match to who you are. It cannot identify the complex combination of reasons *you* might not want to watch a particular movie, nor can it understand why your preference might have changed since the last recommendation you accepted. To do that, it would have to actually understand *causation,* how one thing leads to another in your life, and the machine learning systems of 2018 can't do that very well. They can't know why you wanted to watch that cheesy romantic comedy; just that you're likely to be interested (or not). Causality, this essential ingredient to understanding the way the world works and why, might also be an essential ingredient in getting to the next level of AI systems.

In coming years, we'll learn more about how formidable Unanimous and the notion of swarm intelligence could become, but its idea of empowering people by keeping them in the loop and partnering them with AI systems—embedding human values and wisdom within the AI system itself—hints toward a richer, more valuable, and more potent form of cognition. We have collectively opted to entrust more of our decisions to the recommendations of the global Internet titans, their algorithms, and the vague sort of trust we imbue in them. This might sound like a first step toward a trusting and hyperconnected society, but it also lends itself to the manipulation of power. As François Chollet, an AI and machine learning software engineer at Google, noted in March 2018 Twitter rant: "Facebook can simultaneously measure everything about us, and control the information we consume. When you have access to both perception and action, you're looking at an AI problem." Human minds are "highly vulnerable to simple patterns of social manipulation," Chollet tweeted, and serious dangers lurk in the types of closed loops that allow a company to both observe the state of its "targets" and keep tuning the information it feeds them. (As others on Twitter quickly noted, Google and the other AI titans can wield a similar power.)

Yet, our personal ability to shape our lives, influence those around us, and choose our own pathways remains the most personal source of power. To be sure, how well we exercise that agency depends in part on the quality of information we receive and how well we can process it. But in the end, our cognition and our ability to influence the cognition of others are themselves a form of power. The developing generations of artificial intelligence will lay us bare to the power of other entities to shape our lives, *but it also will help us develop and exercise our own agency and authority.*

We can't go back now. These cognitive machines already shape *human* cognition, consciousness, and choice. They already thrive on our desire to simplify our lives and ease the time pressures we face. This is power we gladly grant in return for convenience or pleasure, and it is also power that is usurped from us without knowledge or consent. If we hope to capitalize on the positive potential these technologies could deliver in the next decade or two, we need to forge a new balance of power in the here and now. It's still our choice, not theirs.

3

Think Symbiosis

The moment embedded itself in the cultural zeitgeist as one of the most memorable technological achievements in modern history, but Patrick Wolff and most of his peers saw it coming for years. "It was a deal, but we knew it was going to happen," he says years later. "It happened, and life went on."

In the late 1980s, as Wolff began ascending the ranks of US chess grandmasters, the idea of a computer beating a human remained the stuff of science fiction—a far off, but foreseeable, milepost that would signal the real arrival of artificial intelligence. The best computer-based game of the day, SARGON, could give most players a serviceable game of chess, but the digital competition was laughable for those with advanced skills. Up until then, players followed magazines and word of mouth to keep up with the latest chess news, tracking their own analyses, observations, and ideas in paper notebooks. But around the time Wolff was winning the US Junior Championship in 1987, professional chess players had started to rely on ChessBase and similar databases, and within a few years every grandmaster carried a laptop with them. They'd get floppy disks with the games played by all the grandmasters, enter their own games, and compile their notes digitally. They could query the database for specific types of openings or defenses.

In 1992, Wolff won his first US Chess Championship, and would take a second title three years later. By then, computer programs had gotten good enough at the game that even the world's best could take them seriously. Grandmasters could still handle their digital opponents with relative ease, but they could augment their analyses of the game by sparring with a decent piece of software. Make a move, leave the computer on overnight to consider its millions of options, and it might make an intriguing move by the morning, Wolff recalls. Still, the path ahead was clear—computer processing power would continue to advance and developers would continue to improve game-playing software. "It was pretty obvious to me and most grandmasters that it was just a question of when," he says.

The moment finally arrived on May 11, 1997, when IBM's Deep Blue beat world chess champion Garry Kasparov on its second attempt. The supercomputer's 3.5 to 2.5 victory made headlines around the world, made Kasparov a household name, and remains an iconic moment in humanity's high-tech achievements to this very day. Yet, life went on in the chess world. Shortly after the turn of the century, commercially available computer systems with dedicated chess-playing software could challenge, and often beat, some of the best chess minds in the world. And over the subsequent ten years or so, the relationship between grandmasters and their digital challengers grew ever more symbiotic.

Then AlphaZero happened, and the chess community still hasn't fully absorbed what it might lead to next, Wolff says. Grandmasters already knew software could process chess better than they could, so they learned how to integrate these powerful digital tools into their workflow to enhance their own game. But those applications essentially worked by executing handcrafted programs that followed these steps: start with constantly updated opening theory, optimize probabilities during game play, and then close out with established endgame procedures. Google DeepMind's AlphaZero developed its expertise upon nothing but the rules of the game, playing against itself and rapidly improving via reinforcement learning. Once trained,

DeepMind developers matched the system up in a hundred games against Stockfish, the best traditional computer program going at the time. Playing with the first-move advantage of the white pieces, AlphaZero won twenty-five and drew the other twenty-five games. Playing with black, Alpha Zero drew forty-seven and won three. Most grandmasters agree that a perfectly played game ends in a draw, so AlphaZero did very well.

To get a sense of how well, consider the Elo rating, which the chess world uses to gauge player quality. It's a four-digit number that currently appears to top out around 3,600. (It used to go to about 2,900, but computers pushed it beyond that level.) Magnus Carlsen, the top-ranked grandmaster who won his first world championship in 2013 and had yet to lose the title at the time this book was written, posted a standard Elo rating around 2,840. Stockfish rated around 3,400, a score that suggests it would be expected to win 97 percent of its games, Wolff explains. AlphaZero's rating nudged past Stockfish in just *four* hours, leveling off right around the 3,500 threshold, according to the information released by DeepMind researchers. "The AlphaZero moment? It was a *moment* for me," Wolff says. "Holy shit. I spent months teaching myself about machine learning and deep neural networks. I needed to understand what the hell was going on."

As he researched the technologies and the ten games about which DeepMind researchers released full details, he realized that AlphaZero, for all its sheer computing and cognitive power, still lacked something innate to human grandmasters: a conceptual understanding of the game. AlphaZero can calculate at four orders of magnitude faster than any human, and it can use that power to generalize across a wide range of potential options across a chess or go board, but it can't conceptualize a position or put it into language. Consider, for example, a game Wolff observed in early 2018, during which one of the players sacrificed his knight in what proved to be an elegant maneuver. To a certain extent, Wolff explains, he could understand the play through pattern recognition, but he also naturally wondered, upon the initial move, why the player would put his knight in a clearly threatened position. "I could understand there was a reason that move was selected," he says, "so I could look at what features changed and (ask) 'Is there a way to use that reason?'" In other words, Wolff was trying to back into what he was seeing, trying to figure out what the larger, conceptual theory behind the move was. What was the chess player seeing that he wasn't yet?

"When I look at the board, I often see an image. That opens up potential theories or ways to play and win." While AlphaZero's neural nets favor certain positions over others, it doesn't appear to gather the same conceptual understanding, which could guide it toward an optimal set of moves. That hasn't limited Alpha Zero's prowess at chess, but it's worth noting because it could have consequences in other domains where a conceptual understanding might produce more holistic solutions to problems. Wolff calls it the difference between raw skill and the sort of imagination that creates conceptual images in the human brain. For the time being, anyway, human grandmasters possess more conceptual knowledge and can imagine a successful theory of winning, even if they're overwhelmed by AlphaZero's skill. Yet, what happens when we combine both?

These days, Wolff says, virtually every grandmaster will train with a computer, integrating it into their routine to help hone their game and develop new concepts—perhaps coming up with a more intricate opening, or contemplating ways to counter novel tactics deployed by opponents. Many of them compete in "advanced chess" or "centaur chess" tournaments, in which human-machine pairs vie against each other. The combination can produce sublime results, says Wolff, who retired from professional chess in 1997 but still enjoys following the top echelons of the human-machine game. "It's like watching gods play," he says. "It's incredible, the quality of chess they play."

SYMBIO-INTELLIGENCE

Most traditional case studies used in business schools and other university settings leave the students who read and discuss them a relatively narrow set of choices. As open-ended a scenario as these exercises set out to simulate, they're inevitably bound by a small number of potential responses. The experiential nature of the exercise typically imparts a deeper understanding of the material and issues at hand—at least better than most lectures—but they fall far short of real-life experience. John Beck and his colleagues haven't created a lifelike educational experience yet, but they've moved a lot closer to that than any case study could.

Beck stumbled across the seeds of a powerful new idea while coauthoring *The Attention Economy* with Thomas Davenport. While researching the book, which argues that companies need to capture, manage, and keep the market's attention, Beck discovered that video games drew and held people's attention far better than almost anything else. He always figured information technology could play a greater role in education, but he never spent much time thinking about how. Somewhere around 2015, something clicked, and Interactive Learning Experiences, or I-L-X, was born. "It's the kind of stuff you wish you could really do in a good case study, but you don't have the time or capacity," Beck says. "Case studies are very narrow . . . but here we have 10^150 possible different outcomes—and even that's a lot less than real life on any decision you're making."

I-L-X uses a video game engine to engage students, but it's not really a video game. It constantly evaluates what a student is learning along the way, blending traditional teaching, video game play, and a kind of scripted entertainment, he explains. The scripted piece of the lesson matters, so the experience might never become an end-to-end AI system, but it already integrates elements of machine learning and could eventually include a range of other AI technologies. But at its core, what makes I-L-X special is the interaction between the human and the experience provided by the machine. It tracks every single move the players make throughout the game, down to the amount of time it takes to make their decisions. "We have a really good sense of what paths people are taking based on what information they're seeing," he says, "and we can change every single element of this game on the fly." By changing the content in a dialogue box in one situation, they can understand how users might change their decisions five or six steps down the road.

For now, much of the expertise and analysis comes from Beck's thirty years of teaching experience, but he already has ideas of how current and future AI integration could increase complexity, enhance the sociological and emotional context of the stories, and interject a range of current events. For example, in one class back in 1998, Beck first used a current event—the crash of SwissAir Flight 111 out of New York City—in an educational "war game" about terrorism and the airline industry. Now, he revels in the idea of using AI technologies to introduce current events into games in real time, bringing a deeper context to the students' interactive experience.

Better yet, he expects to use more cognitive computing to enhance the ways I-L-X already injects different types of learning experiences into the scenarios it presents. Beck says that once students are stimulated by the interactive experience, providing them a range of learning styles helps them understand and retain knowledge far better than identifying one that suits them and going deep with it. "We can take somebody out of their comfort zone, their preferred thinking and learning style, and hit them with a situation that requires a different one," he says. (In one game he's designing there are eight to choose from.) This can be especially powerful when I-L-X interlaces business content of the case with a personal relationship narrative, just like in real life. "We can make students feel really uncomfortable and provoke them both intellectually and emotionally, and that's when the most significant learning happens."

So, rather than the intuitive idea of just using AI to figure out the best technique to impart a lesson to each individual, I-L-X could also use it to, say, find the best order of a wide variety of techniques to deliver the same information. And that, Beck says, often depends on the context in which the students find themselves. Inject one methodology or another at the exact right time, based on what the game senses from the participant, and thus deliver the best and deepest educational experience.

The power to change the context in a learning experience through a dynamic interaction between humans, cognitive machines, and the real external environment provides a tremendous opportunity to enhance education around the world. Enio Ohmaye, the chief experience officer of EF Education First and former chief scientist of Apple Japan, has put that kind of smart contextual simulation to work to help young Chinese students learn English. Ohmaye's young son is growing up in an environment where he's exposed to four languages—Portuguese, Japanese, English, and German. "There's not necessarily anything unique about his brain physiologically speaking," Ohmaye says; by which he means that most other kids in the same environment would pick up the same skills. His son just absorbs them from the surroundings—the Portuguese from his father, the Japanese from his mother, and the German and English from his school. Few youngsters in China have the same experience. "It's not about tech," Ohmaye says, "it's about exposure to language in a way that is relevant, meaningful, and effective for them to learn the language."

And that's where the interaction of AI systems, the young human student, and his environment comes into play. EF Education First develops systems that bridge between the physical and digital worlds. Using a variety of machine learning and image recognition techniques, in early 2018 the firm introduced a program in China that's designed to more deeply permeate the life of the young user, integrating the English lessons with the human elements of his or her young life. The firm is patenting a little robotic pet that recognizes what's going on in the student's environment and adapts its English lessons to fit the situation. So, depending on the time of day, past behaviors, and the current situation, it might start interacting with the student by using words associated with bedtime routines—phrases associated with brushing teeth, saying goodnight, singing lullabies, or reading bedtime stories. "We begin to permeate the life of kids with English that's really relevant," Ohmaye explains. "The beauty of it is we can, in the process, make it a little fun, and we can involve the parents just by virtue of the fact we're helping the parents create behaviors and structures they (the kids) need to abide by."

Yet, it's not sufficient to just have the toy. EF Education First supplements this with programs that forge a symbiotic ecosystem between the pet, the child, the parents, and the environment. Parents, grandparents, and teachers all play a role in the interactions, often as drivers of content. It's a combination of device, social support, and a human and machine ecosystem encompassing that community. The back-office AI system can track what kids listen to and what teachers say, and by doing this across thousands of students it can identify better techniques. That's what gets EF Education First past the roadblocks of most online education programs, which get plenty of people to sign up but few to complete. "This marriage of physical and digital and the computational power of AI allows us to create a much more immersive experience," Ohmaye says. "You send a 16-year-old to France, and they come back transformed. That's what made me fall in love with this company, delivering these transformative experiences."

That sort of symbiotic relationship between artificial, human, and other types of natural intelligence can unlock incredible ways to enhance the capacity of humanity and environment around us. Yet, as humans, especially in Western cultures, we tend to order our existence in terms of hierarchies—a struggle to ascend the food chain, climb the corporate ladder, or remain atop the evolutionary pyramid. We apply the same conceit to artificial intelligence, constantly ranking it against our own intellect and fretting about when it exceeds our capabilities. Humans do this naturally when trying to figure out how to relate to something new, says Genevieve Bell, anthropologist and professor of computer science at Australia National University and senior fellow at Intel's New Technologies Group. Bell likens it to the Buddhist concept of "dependent co-emergence." By comparing artificial intelligence with ourselves, she says, our understanding of each is clarified by the other.

Still, Kevin Kelly, the founding executive editor of *Wired* magazine, would appreciate it if we stopped thinking that way, at least in terms of intelligence. "Intelligence is not a single dimension," Kelly says in an April 2017 column for his publication, "so 'smarter than humans' is a meaningless concept."[*] Rather, he argues, the world is full of a wondrous array of intelligences, each having evolved over time to their currently refined state. Even without neurons, colonies of certain slime molds can solve mazes, balance their collective diet, and escape from traps.[†] Bees exhibit remarkably complex problem-solving capabilities through the collective intelligence of the hive. Whales have significant social intelligence. Humans generalize concepts and imagine new things in ways no other animals can match. And AI systems can process complex mathematical and memory feats that no human can ever hope to accomplish with their gray matter alone. Kelly likens this spectrum of intelligences to symphonies with myriad instruments that "vary not only in loudness, but pitch, melody, color, tempo, and so on. We could think of them as an ecosystem."

Stuart Russell, who leads the Center for Human-Compatible Artificial Intelligence (CHAI) at the University of California, Berkeley, finds Kelly's argument unconvincing. Human intelligence is complex,

---

[*] Kevin Kelly, "The Myth of the Superhuman AI," *Wired* (April 25, 2017).

[†] Ed Yong, "A Brainless Slime That Shares Memories by Fusing," *The Atlantic*, Dec. 21, 2016.

Russell says, but machines might eventually surpass it in some general manner—for example, by exhibiting the ability to perform almost all human professions as well as people can. Russell and Peter Norvig published the first edition of their now-standard textbook, *Artificial Intelligence: A Modern Approach*, in 1995.* Two decades later, and after thirty years of AI research, Russell began to wonder: "What if we succeed?" He and his colleagues at CHAI aim to reorient the field, essentially working to reestablish it with safety built into its foundations. They want to create what they call "provably beneficial systems" to make sure any new form of AI, regardless of its mundane or superintelligent capabilities, is created with the ideals of human safety and benefit at its core.

Our fascination with and fear of artificial intelligence might stem in part from our hierarchical thinking of the world and the idea of a superintelligence subjugating us. But it also arises from the fact that we long have thought of and modeled AI systems after our own human brains—that the super in that intelligence could mean our brains raised to an exponential power. Yet, despite the fact our particularly human capabilities have enabled us to dominate other species to the degree that we now live in what many call the Anthropocene Era—a time dominated by human decisions and technologies rather than other natural forces—we can't really claim to have arrived in this advantageous position with a level of responsibility and care that's commensurate with the power we currently wield. For most of our history, humanity behaved as a ruthless, adaptive dominator of the natural world around us. Creating tools is one of our signature adaptive techniques; now, we find ourselves with a tool that thinks and talks back.

In the 1970s and 1980s, when nature "talked back" and we recognized some of the irreversible damage we were causing, environmentalism represented an assertion of the power of nature over humanity. Yet neither environmentalism nor humanism provide a satisfying explanation for the evolution of cognitive machines in the 2010s and 2020s. Intelligence is neither the prerogative of humans nor nature. It is an evolutionary act of cross-fertilization, with different types of intelligences interacting and mutating into new types of intelligence altogether. Over

---

* Stuart Russell and Peter Norvig, *Artificial Intelligence: A Modern Approach*, 3rd ed. (New York: Prentice-Hall, 2009).

the last 2.8 million years or so, humans have used nature's resources, domesticated many animals, and impacted the evolution of a myriad of others, taking influence over different forms of intelligence in nature. Over the last seventy years, humans have used nature's resources and manipulated physics to make ever smarter computers. We evolve, we interfere and meddle, we create, and we force other forms of intelligence to coevolve with us. So, why would machine intelligence be any different? Whether legitimately or not, we have come to view artificial intelligence as threat, rather than a new intelligence with which we can partner, enhance our ecosystems, and enrich our lives. Like the chess grandmasters using computers to raise their games, and like those same experts trying to figure out what AlphaZero will mean for the game they love, we currently struggle to embrace a new world of *symbiotic intelligence*.

Symbio-intelligence represents a cohabitation and integration of multiple forms of intelligence—human, nature, and computational machine—into a coemergent and cocreative partnership that benefits all sides. It produces benefits to each contributing entity that it could not enjoy on its own, such that the new partnership exceeds the sum of its parts. Ken Goldberg, an artist and roboticist who leads a research lab at UC Berkeley, proposes the similar idea of "Multiplicity" as an inclusive alternative to the Singularity, or the hypothetical point in time when computers surpass human intelligence. Goldberg's notion of Multiplicity emphasizes the potential for AI to diversify human thought, not replace it.* "The important question is not when machines will surpass human intelligence, but how humans can work together with them in new ways," Goldberg says in a *Wall Street Journal* op-ed. "Multiplicity is collaborative instead of combative. Rather than discourage the human workers of the world, this new frontier has the potential to empower them."

Even within the cognitive realm, human, natural, and machine intelligences display distinct powers thanks to their unique evolutionary pathways. Human brains, for example, process data and handle our complex balance of bodily operations with an unprecedented energy efficiency. Millions of years of evolution and "genetic intelligence" have led to an incredibly intricate human ability to manipulate objects with our hands, something that will take massive amounts of computing power and research to replicate in robots. And that

---

*   Ken Goldberg, "The Robot-Human Alliance," *Wall Street Journal* (June 11, 2017).

doesn't even begin to scratch the breadth of human processing and function. For example, consider a professional tennis player reacting to an opponent's 150-mile-per-hour serve. As Stanford humanities, sciences, and neurobiology professor Liqun Luo explains, the thousands of interconnections in the player's brain allow him to spot and process the flight of the ball, almost immediately identify its trajectory, and begin moving his legs, torso, shoulders, elbows, wrist, and hand into position to return the serve, all simultaneously and in concert.[*] "This massively parallel strategy is possible because each neuron collects input from and sends output to many other neurons—on the order of 1,000 on average for both input and output for a mammalian neuron," Luo writes. "By contrast, each [computer] transistor has only three nodes for input and output all together." This enables the considerable multifunctional dexterity of the human body that is so hard to achieve for most animals and robots to date.

Yet, computers already surpass human performance in comparatively new endeavors, such as chess and Go, despite the remarkable efficiency of the gray matter in our heads. The human brain uses about twenty watts of energy, roughly double the amount of power delivered by an iPad's minicharger, explains Sean Gourley, the CEO of Primer AI, which builds machines that can read and write.[††] Today's advanced computers consume far more energy, but the sheer speed of their serial, step-by-step processing abilities—combined with some parallel capabilities—allow them to easily outperform humans on a range of tasks, such as image processing, decision-making, and text recognition. Maybe the most severe limitation humans face is in the realm of dimensionality. We struggle with just three dimensions whereas computers routinely work with thousands of them. But for so many applications, a computer's massive advantage in processing speed makes all the difference necessary. "That's why we let computers trade on Wall Street today, rather than humans," says Gourley, who has advised the US government on the mathematics of war, among other things. "Trading is a narrow task with just a few vectors for decision making, mostly driven by the speed

---

[*]   Liqun Luo, "Why Is the Human Brain So Efficient?," *Nautilus*, April 12, 2018.

[†]   Interview with the authors in San Francisco, January 29, 2018

[‡]   Christina Bonnington, "Choose the Right Charger and Power Your Gadgets Properly," *Wired*, Dec. 18, 2013.

of processing pure economic-financial metrics. That is much better suited to a narrow intelligence of a computer, whereas the human brain is a bit more holistic but a lot slower."

By contrast, human brains are easily diverted by predispositions, distractions, and confirmation bias, when we value information that confirms our beliefs and miss critical details in what we observe. The same focus mechanisms that block out sensory distraction so we can make simple decisions also cause most of us to miss the gorilla that walks through the basketball game.* And yet, this same capability to think about facts from different angles and take a little longer to process decisions may equip humans with a better "provocateur intelligence," Gourley says. We can critically reflect on decisions and put them into the larger context of societal needs, multiple stakeholders, and developments across multiple domains, not just economic and financial ones.

Machine algorithms don't assess the social consequences of a merger and acquisition activity, such as layoffs, and the potential externalities that these bring to society, unless they are specifically programmed to do so. The human brain can and often does, in part because we possess an empathetic element in our intelligence that leads to moral considerations and deliberation across a broader set of different factors. Developers concerned with creating effective and efficient machine algorithms to trade and maximize the value of our equities in our retirement portfolios might view empathy as a risk to their mission, a consideration that slows down critical decisions and muddles the picture.

Building an environment of symbio-intelligence requires the ability to recognize the good and the bad parts of various types of intelligence so we can merge them into an optimized partnership. It relies on a deeper sense of trust in the potential of AI and a willingness to accept that human intelligence can be enhanced by the same intellectual powers we once reserved only for science fiction. Yet, around the world, we already are seeing powerful new applications of advanced

---

* With this test, Chris Chabris and Daniel Simons showed that humans who focus too hard on a narrow task will miss critical elements in their field of observation. The test shows two groups of people playing basketball, and it asks the audience to count how often the ball gets passed between members of one team. Focused on counting, most observers miss the man in a gorilla suit who moonwalks through the scene. (Daniel Simons, *The Invisible Gorilla,* (New York: Harmony, 2010); and https://en.wikipedia.org/wiki/The_Invisible_Gorilla)

technologies that automate not just human thinking, but virtually every facet of our identity, our experience, and our well-being.

STAND-UP COMEDY, RUGBY, AND THE INEXTRICABLE RELATIONSHIP OF MIND, BODY, AND ENVIRONMENT

Kevin Kelly likens this notion of symbio-intelligent relationships to instruments in a symphony. John Neal sees it in the elite athletes he helps train. Neal, the head of coach development at the England Cricket Board and a professor of sports business performance at Hult Ashridge Executive Education, serves as a performance coach for various English teams and individuals, including the royal household. Having left behind traditional psychology for a more symbiotic approach that merges emotional insights with physiological data, his goal is to bring out what he calls "flow" in his coaches and athletes, and he does so by recognizing the inseparable relationships between mind, body, and environment. It starts with the measurement of physiological and neurological signals, determining how people learn, reflect, recover, and perform. The brain and body, Neal says, provide an honest accounting of their state. An athlete might say one thing, but the data are "absolutely binary, like an autistic response," he says, meaning black and white, lacking nuance and empathy.*

That autonomic data becomes a powerful instrument with which to gear training programs for the fractional improvements that often decide contests played at the most elite levels. But those athletes, whether teams or individuals, don't play in a vacuum. Neal and his colleagues also train athletes to prepare for the types of environments that might keep them from peak performance. When athletes feel confident, he explains, they go into that state of "flow." The game moves in slow motion. They anticipate what will happen next, staying ahead of their opponents. They typically feel great and perform well. However, if they tip out of flow into a state of challenge or threat, Neal says, actual blood flow to the brain starts to change, and they begin to spend more neurological and physiological effort on strategy and decision-making. They fall back to old patterns.

---

* Interview with the authors via video conference, November 14, 2018

Coaches can see it happening in their athletes, sensing not just a drop off in performance but a change in the baseline state of a player. The best coaches know when to intervene and how they might snap individuals back into flow. Sir Clive Woodward and his fellow coaches noticed it during the 2003 Rugby World Championship semifinal, Neal recalls. Jonny Wilkinson's play had dropped off, but one of the coaches realized he'd started moving differently, had changed his usual body language, and appeared almost in a panicked state. Many coaches would've taken Wilkinson off the field, but in a quick huddle—maybe forty seconds or so, Neal says—the coaches collectively decided that they needed to keep him in and preserve his confidence and mindset for future matches. So, they sent Matt Catt in, substituting him for another player. On his way on, Catt ran by Wilkinson, slapped him on the backside, said something upbeat to him and laughed. It was enough to snap Wilkinson back into the game, and he went on to contribute to the winning score. "It was the most remarkable piece of coaching I've ever seen," Neal said. "There was a lot of debate between the coaches, and it was heated, but the decision was made in about forty seconds. It was a perfect example of intuition, intelligence, and cognition coming together in a very short period, and it was all on TV."

Both the "intelligent intuition" of the coaches and Wilkinson's snap back into a state of flow illustrate how the subtlest changes in environment can recalibrate the performance of mind and body. If an athlete doesn't feel comfortable in an environment, he or she will switch back to the emotional cortex and back to self-awareness, perceiving the world primarily through a lens of negative experiences and a threatened state, Neal explains. We can't change that natural reaction, but we can learn to identify the situations that activate this response and, with training, refocus those thought patterns through a physical activity—say, stand-up comedy.

Yes, an occasional piece of Neal's training regimen takes place not in the gym or on the field, but in a comedy club. For Olympians who have never been on that big a stage, Neal and his colleagues create a similar environment by giving athletes an hour to prepare a comedy routine. They're terrified; the idea of performing comedy for a roomful of people gives them a real sense of fear, but it also provides the coaches an opportunity to help them relax when put in that context. They begin to realize they can go into a threatened state, but they don't need to fail. Over the course of their training, Neal says, they eventually get to a point where they want to do standup because they enjoy it. "And then we can use heart and other physiological data to show them, in hard data, how their performance is improving," he says. "Heart rate variance data shows how their physiological response is changing, and once they see it they believe it even more."

This intricate, complex, and intractable relationship between our minds, bodies, and the world around us make up human reality. Alan Jasanoff, a professor of biological engineering at the Massachusetts Institute of Technology, calls this the "cerebral mystique"—the idea that the connections between stimuli, emotions, and cognition are not discrete and separable functions.* Our brain should not be viewed as computer analog, Jasanoff suggests, largely because what makes us human is the complex interplay between physical sensations, emotions, and cognition. Our bodies are part of cognition, so our physical environment, social context, and life experiences shape our self-perception and identity. From the early stages of our childhood through the last moments of our existence, we receive and integrate feedback from the people and institutions around us, and those physical and emotional stimuli become part of our cognition, our experience, and ultimately our values.

Our environments teach us about right and wrong behaviors, good and bad performance, and standards of beauty and aesthetics. It shapes who we are and how we move through the world. Social control mechanisms in our communities and organizations signal to us the acceptable norms for interaction. And in all these processes, our own notion of our identity meets, melds, and clashes with the perceptions of others we encounter. We shape and reshape our sense of self through a lifelong process, struggling to balance our power, values, and trust with one another and the institutions with which we interact.

In the past, we retained a certain share of power over how we represented ourselves in that public sphere, but not always for the better. Politicians would portray themselves as great and visionary leaders, while voters sought to find out the truth about the men and women who wished to lead us—their integrity and flaws helping us

---

* Alan Jasanoff, *The Biological Mind*, (New York: Basic Books, 2018).

gauge whether they were fit for office. But we also accepted a certain amount of ambiguity in the formation of character and public identity, especially for candidates we supported. We chose to forgive their errors or justify their successes as outcomes of complex, multifactor pictures. And we accepted them as part of a subjective narrative, one that appeals to us even if its parts were less than pure truth.

However, the prevalence of personal data and artificial intelligence will change how we craft notions of self-image and self-awareness in the years to come. Smart machines will make us more objectively measurable, much like an athlete whose physiological, neurological, and environmental signals are gathered, read, and acted upon by coaches such as Neal. The emerging algorithms, deep neural networks, and other sophisticated AI systems can process dozens of data streams from a multitude of areas in our lives, assessing our overall performance, profile, or character traits. They might not always be right. Data can be biased, and people can change, but already these systems capture intricate facets of our existence and our environment that we don't have the type of intelligence to fully process.

Consider, for example, the ability of IBM Watson to conduct personality assessments based on your Twitter feeds, a capability it designed with the help of a panel of psychologists.[*] While the system, at the time of this writing, still had some kinks to work out, such as distinguishing between an original post and a retweet, the algorithms could conduct a text and sentiment analysis from the feeds and deduce personality profiles. Already, human resources departments have deployed similar systems to detect happiness or discontent among workers who answer open-ended questions on annual employee surveys, according to the Society for Human Resource Management.[†] It's not a far stretch to imagine such systems applied across all of a person's social media feeds and data streams, and from that picture Watson might assign labels such as introverted or extroverted, passionate or aggressive, open-minded or closed to the ideas of others. It's an astoundingly simple step toward providing us an external check, a mirror to see how other people might perceive us.

---

[*] IBM Watson Personality Insights (http://watson-pi-twitter-demo.mybluemix.net)

[†] Dave Zielinski, *Artificial Intelligence and Employee Feedback*, Society for Human Resource Management, May 15, 2017.

This raises plenty of opportunities and risks, not the least of which is how accurate a view of a person's true persona such a system might provide. We heavily self-curate our portrayal of ourselves on social media—in part because most of our everyday actions and communications haven't been publicly digitized yet, but also because we present ourselves differently from what we really are. At one point, while working as a vice president and senior fellow at Intel, Genevieve Bell and her team suggested the idea of integrating social media into set-top television boxes, so people could quickly and easily share what they were watching rather than typing it into their phones or laptops. In testing, users revolted because they regularly lied about what they were really doing, says Bell, who's developing the new Autonomy, Agency and Assurance (3A) Institute at Australian National University. "Human beings spend most of their time lying, not in the big nasty sinful ways, but the ones that keep the peace and keep things ticking," she says. "Systems don't know how to do that, and humans don't know how to live in a world without that. . . . If everybody just lived out their true impulses without control screens or masks, the world would be a terrible place."

Of course, as almost anyone on Twitter knows, people can still hide behind incomplete or fragmented digital identities and use that as a shield to unleash their basest impulses. Digital trolling has become commonplace over the past decade, spawning levels of vitriol that rarely occur in face-to-face interactions. The digital stalking, intimidation, and hate won't dissipate any time soon, especially not in societies and on platforms that favor free speech. Yet, as the digitization of our physical world increases, we might see a ratcheting up of the tension between digital and real-life personas. Data streams will become more holistic, multifaceted, and comprehensive. Arguably, then, the picture of who we are will get triangulated and checked and veritable. How will this impact the ways we perceive ourselves and the ways others perceive us? How will this new blend of data-based assessments and human masks shape our consciousness? How much control will we still have over our self-awareness and self-image when we are constantly forced to look into the mirror of data-driven authenticity? When digitally objective historical data tells one story, but our aspirations tell another, how much influence can we exert over how we evolve and who we become? Might the evolving society then become more rational, honest, and less prone to egos and misdirection, but also no longer a place for free self-definition?

Already, thousands of people rely on the algorithmic matchmaking "wisdom" of online dating sites to find partners. At many of them, AI-driven platforms help cut through the self-promotion and clutter in personal profiles to hone in on a more likely match. These sorts of systems might help us adjust our expectations of ourselves and others, so we're disappointed less often and develop happier relationships. We might find partners that better fit our true nature, rather than hooking up with the people our egos and social conditioning guide us toward. Given the mediocre success rate of the self-guided approach to romantic love, a little help from an objective source might not be a bad thing, as those who grow up in arranged-marriage cultures sometimes argue.

Yet, all this assumes that the AI platforms involved in the process get it right enough to build trust in the system and, by extension, between us as individual citizens of our communities. These systems and the people who develop them have yet to fully address concerns about bad data, human and algorithmic biases, and the tremendous complexity and nonlinearity of the human personality. To computer scientists, it is more like a piece of art than a piece of engineering. But their current training allows them to treat it only as the latter. What if the geeks get us wrong? What if the psychologists involved apply monocultural lenses to cross-cultural identity issues?

Additionally, even when done perfectly from a code and data perspective, AI can only measure what it sees. Much of who we think we are lies in ideas we never speak, or we keep them in such finely nuanced patterns that AI can't detect it. As Genevieve Bell notes, a small dose of self-deception can boost self-esteem, smooth our relationships with the outside world, and carry individuals and communities a long way toward success. But then, if the right balance of reality and delusion is key, how will AI capture that balance and the almost intangible timing of pivoting from one to another? That could backfire if the AI doesn't get the fine lines.

Ultimately, a beneficially symbiotic relationship between our minds, bodies, and the environment—as well as with other forms of natural and man-made intelligence—relies on the AI agent balancing the tension along those tenuous lines. As humans, we need to maintain the power to make sure this high-wire act goes well. Otherwise our trust in each other will erode and our trust in AI will die before it can bear fruit.

EMOTIONAL INTELLIGENCE AND TRUST

Dr. Alison Darcy and her team refer to it as "he," fitting given just how intimate and attached "his" users can become. This character they refer to is Woebot, an AI-powered chatbot that provides psychological therapy services for its users. At first glance, it harkens back to the conversational banter users could have with the famous ELIZA, one of the earliest natural language processing computer programs developed in the mid-1960s. Woebot goes considerably deeper in its engagement, though, and many users have developed a close bond with it. Their attachment might stem in part from just how lonely they say they are, a fact that surprised Darcy, a clinical psychologist at Stanford University and Woebot Labs' founder and CEO. Even in social situations, users, especially young adults, report that they feel lonely. One colleague noted how often her adult patients would describe an empty and hollow feeling. They started seeing that in Woebot's generally younger clientele, who welcomed the five- to ten-minute moment of self-reflection that the mobile phone app could provide.

Yet, Darcy stresses, they did not create Woebot to replace a professional therapist. The app merely checks in once a day with a lighthearted and friendly interaction, complete with emojis and banter. If you do say you're feeling down, it will drop out of its jovial mode, which simulates empathy, and bring you back into the cognitive restructuring that's at the core of cognitive behavioral therapy (CBT). Loosely speaking, CBT is based on the idea that what upsets someone is not the events in their world, but their perception of themselves and what that means about them. Thus, treating the root cause of those anxieties, a process called cognitive restructuring, requires active participation on the part of the patient. Although research shows it to be effective, only roughly half of psychotherapists employ the techniques, and fewer do it well. A big part of the problem is keeping patients engaged, doing the homework they need to do to "overwrite difficult thoughts with rational ones. It's a fair amount of work for the patient,"

Darcy says, "but it really pays off and that's empirically proven, so it does make sense to have a coach reminding you to do the work."

Woebot can only approximate a good, long-term CBT intervention and it's not designed to replace the human counselor, but two things make it especially useful. For one, it's available at any time, including for that midnight panic attack on Sunday. While not intended to provide emergency services, it can recognize dire situations and trigger suggestions to other services, including an app that has an evidence-based track record of reducing suicidal behavior. In fact, Woebot Labs, which received an $8 million Series A funding round in March 2018, takes patient control of the app, their interactions with it, and their data so seriously that they allow users to turn off the alert words that usually trigger the app to challenge their suicidal thinking.

The second element that makes Woebot so useful is the fact that "he" builds unusually strong bonds with "his" users, forging a working alliance with them and keeping them engaged with their therapy. While the folks at the company refer to Woebot as "he" because they're referring to the character it portrays, Darcy says, users also refer to it in remarkably personal terms. "They used relational adjectives," she says. "They called it a friend." In fact, she explains, one Stanford fellow's study of Woebot's "working alliance" measures found that it did not map with what he'd previously seen in computer-human interactions, nor did it map with human-to-human interactions. For her part, Darcy suggests it occupies a space in that gray area where people suspend disbelief, much in the same way we encourage kids to talk to their Elmo doll or pretend their toys are real. Yet, Woebot's developers specifically avoid making it more lifelike or realistic. They want it to remain a distinct entity, clearly a therapeutic chatbot—something removed from simple friendship the way a psychiatrist or doctor maintains a professional demeanor and code of conduct. "A really good CBT therapist should facilitate someone else's process," she says, "not become part of it."

Similar types of trusting, empathetic, and symbio-intelligent relationships between bots and their users have proven remarkably effective in addressing a variety of mental health issues. Other psychotherapy bots, like Sim Sensei or ELIZA, have effectively treated US soldiers suffering from post-traumatic stress disorder (PTSD), providing them an effective, base-level therapy while also providing a remove from a human therapist, with whom they might hesitate to interact. With a watchful human eye, these AI-backed systems have penetrated something as innately personal and human as our mental health—and because of the regrettable stigma still attached to most mental health services, they might work better than human interaction in many cases.

A young start-up that emerged from Carnegie Mellon University hopes the same symbiotic relationship between AI and humans can help address the opioid crisis that sunk its teeth into American society in 2017. In some ways, Behaivior moves a half step further from Woebot, more actively engaging in actions designed to nudge recovering addicts away from potential relapses. But it also takes a cue from Darcy and Co., making sure the control of the app remains in the hands of the user—something critical to establish and keep trust. Behaivior works in concert with wearable devices, such as Fitbits and mobile phones, to measure a range of factors about the recovering opioid addicts who use it. It draws on everything from geolocation to stress and other physiological metrics to identify when a user might be heading for a relapse. While still in initial testing when cofounder Ellie Gordon spoke with us, the AI-backed system detects nuanced patterns in activity of individuals and across the range of its user base. When a factor or combination of factors signals the possibility of an imminent relapse—say, signs of stress and a return to the place where a person used to buy heroin—it triggers an intervention predefined by the user. For some, Gordon says, it's a picture of them in the throes of drugs, looking gaunt and unhealthy. For others, it's music. Many recovering parents select pictures of or messages from their kids. And sometimes they set up a connection to a sponsor from a twelve-step recovery program.

The systems can pick up on some intriguing signals, such as cigarette usage. If users start smoking more often, Gordon explains, they're likely regressing toward a high-craving state. Through sensors on a Fitbit or a similar device, Behaivior can measure the movement of arms and hands when users smoke. It's not infallible, of course. A person might smoke with the hand not wearing the Fitbit, so collecting a wide array of less-than-obvious signals is critical. Recovering addicts relapse far more often than most outsiders realize, Gordon says. It's not unheard of for addicts to overdose, go to the emergency room, get treated and discharged, call a dealer, and immediately overdose in the bathroom of the very hospital that just treated them. Because of the relapse rate—six or seven falls aren't uncommon before treatment sticks—Behaivior has started to gain traction with treatment centers. The founders hope to eventually extend that interest to insurance companies, which could save money on expensive coverage of the multiple treatments, especially in emergency rooms. For now, though, Behaivior remains in the development phase, with much to research, test, and prove. But its ideas have made it a finalist in the $5 million IBM Watson AI XPRIZE competition. The start-up's AI systems help identify the nuanced behavioral patterns and data that signal potential relapses, but eventually they could use such technologies to improve interventions in real time, learning on the fly to help pull people back from a potential relapse.

Both Behaivior and Woebot intercede in some of the most intimate and "human" aspects of our lives—our mental health and our most powerful cravings. Those interactions can work only on a foundation of trust, so both firms take pains to make sure the user has control over the experience. But both firms also prove that these systems can dredge up ever deeper insights into ourselves—our cognition, our intelligence, and even our emotion. It's easy enough to dismiss the idea of a machine understanding our emotions, says Pamela Pavliscak, the founder of Change Sciences and a faculty member at the Pratt Institute. However, Pavliscak, who studies the human-machine relationship using a combination of ethnography and data science, notes that humans don't do an especially good job at identifying emotion in others, either. "The more I looked into emotion AI," she says, "the more I saw potential for it to help us humans learn more about emotions we don't know."

Currently, technology only skates on the surface of emotion, partly because our emotions don't always manifest themselves in physically measurable ways. We portray our state of mind in an array of signals. Some are clearly recognizable, such as speech or body language; some only send subconscious hints of what we feel; and others never get expressed at all. Furthermore, emotion has a cultural component, accruing layers of meaning with the accretion of memory and within specific contexts over time. The technology still has its own limits, as well. "As someone who gets motion sickness," Pavliscak says,

"I still wish GPS knew not to give me winding route." In fact, one might imagine an AI-powered app that could provide more than just a preferred physical route. For example, Pavliscak imagines one that could sense the emotional climate of a particular roadway, giving you a choice between a faster route whose drivers emit an angry vibe, or another option that takes a few more minutes but has a peaceful, calm sensibility about it.

An AI system's ability to process that vastly varied data could help us learn about emotion in new ways, Pavliscak suggests, and it could be manipulated to exploit us. Either way, it will influence our behavior, and often without us really knowing why. Inside that block box, how does the machine really regard us? Can we create a machine the steps beyond intelligence into a more humanlike meta-reflection about us, some kind of quasi-consciousness? And if so, could we ever know what it thinks of us? Would we have a right to know, or would our access to that be limited?

HOW I LEARNED TO STOP WORRYING
AND TO LOVE THE MACHINE

The immersive experience that keeps students so engaged with John Beck's Interactive Learning Experience spins off mounds of data that his firm can use to improve the program. The developers of video games, the initial inspiration for Beck's role-playing education platform, harvest the same data and insights. Users generate tremendous amounts of data just by playing, disclosing their location, employing strategies, winning and losing, and paying for the experience. Developers track user decisions and the context within which they make them. They test various in-game experiences to study player reactions. And they use the combination to move closer to the Holy Grail of entertainment—a gaming experience tailored to an individual's desires, current situation, and financial capacity. Players get a game that can fill and fulfill their day, speak to their cognitive ability to absorb and enjoy it, and take their minds off the burdens of life.

That loop of deep engagement, data generation, testing, and measuring reactions to improve the experience—and thus heighten engagement and/or spending—has spawned an entire cottage industry, and Bill Grosso sits in the middle of it. He took a circuitous route to get there, twice dodging the siren song of academia. The first time, as a young mathematician in his mid-twenties, he decided a comfortable professorial career at a middle American college wouldn't do. He hopped over to the software start-up world, where he became intrigued by what he might do in AI. A few years of research at Stanford, and he found himself right back on the academic track again. So, he skipped back out and led a variety of start-ups for the next decade or so.

By 2012, Grosso noticed an "enormous revolution" emerging. "I was helping run a financial payment processing company with huge data," he says, "and I realized almost everything we took for granted about our behavior was going to come into question." He'd recognized three major trends emerging: ubiquitous mobile phones pumping out data; the cloud providing high power computing for low cost; and increasingly robust machine learning. "I'd realized the science of measuring behavior just became possible," he says. "You can measure, can run an experiment and see how people subtly change their behavior, and you can do it all from infrastructure in the cloud."

Grosso launched a start-up called Scientific Revenue to help clients increase in-app purchases with a dynamic pricing engine for mobile games. It works entirely with digital goods, like the types of small-value purchases popular in video and online games. If a game has millions of players, and you can capture fine-grain measurements on their play, transactions, and context, then the company can change prices to optimize sales. Offering a player a new weapon or bag of gold at an opportune time sounds simple enough, but it gets much deeper than that, Grosso explains. "We're collecting somewhere between 750 and 1,000 facts about you every time you play the game—what device, battery level, source level, time, etc.," Grosso says. "Then, we also capture the information about what you do in the game itself. Did you spend coin? Did you level up or try to level up? How long [was] this session going?"

The depth of information allows Scientific Revenue to create what Grosso calls "an insanely detailed graph of your behavior." But he insists the firm doesn't care about individual data—and, in fact, the firm never stores personally identifiable information. What really matters, he says, is the patterns and irregularities they can draw out of the data of, say, 500,000 players. That allows game companies to shift prices for larger groups of players who might be easily enticed to pay for a virtual item or upgrade. And after that, they can measure the reaction from a granular level up to a collective level, to better understand the most potent causes of consumer decisions. So, players in the game for seventy-three minutes a day—a time that corresponds to a certain level of addiction—might get higher cost, bundled offerings rather than a low-price, one-off item designed for a novice.

But what sort of concerns does Grosso have about gathering that much data on so many individuals? Enough, he says, to have a blanket rule to not store any data that might allow an expert hacker to draw out individual information. If someone stole the firm's data, they could learn a tremendous amount about gaming and consumer behavior, but nothing that shows John Doe playing at 11:32 P.M. on Tuesday night at an all-night coffee shop in Chicago. "I'm not claiming we're super virtuous," he says, "but we don't store ethnicity or legal categorizations, and we don't store any information that can be used to contact you."

Grosso acknowledges the broader risks. As companies classify customers into more and more nuanced buckets, they could start to microslice society with differential pricing on anything for anybody. The possibility of discrimination based on a variety of factors, whether critical or mundane, becomes increasingly easy and likely. "We're already doing this though," Grosso says. We love that Amazon suggests just the right product at just the right price, but we freak out when Target starts identifying a young woman's pregnancy before her father is even aware of it.* There's no nirvana of fairness as we are assessed minute by minute through the triangulation of our data streams. Companies always will create unequal classes of customers, like airlines providing preferential treatment to frequent, often more affluent, flyers. However, taken to extremes those discriminations could splinter communities because no one rides in the same boat anymore. The glue between us, the loyalties and allegiances of being in the same situations, walking through life facing similar challenges, may be lessened as we get microsegmented in pseudoscientific ways based on granular differences.

As the Internet of Things (IoT) becomes more pervasive in the coming years, with more sensors and greater processing power in

---

* Kashmir Hill, "How Target Figured Out A Teen Girl Was Pregnant Before Her Father Did," *Forbes*, Feb. 16, 2012.

virtually every kind of device—all streaming our data back to the cloud—the corporations who operate and control the entire network will have an intimate view of our "life patterns." The AI systems in these networks will make more microdecisions for us. They'll pick a route to our destination, automatically adjust our calendars, book restaurant reservations, restock our refrigerators, and pick out just the right anniversary gift for our spouse. The air conditioning system in a future BMW will let the thermostat in your home know that you've been feeling a bit chilly today, so your home is at the temperature you prefer. Your toilet might let your fridge know it's time to order more vegetables so you get the right dose of fiber or vitamins. And your phone's facial and voice recognition software will instruct your home stereo the right tune to play for the mood you're in.

All of this could make for more enjoyable and, hopefully, less transactional lives. But like simple Amazon and Target contrast above, it cuts both ways. The machine will need to learn which decisions one would happily outsource, which decisions they prefer to retain, and how that differs from one person to the next. And as we experiment, we might hit some wobbly stretches of human-machine interaction. John Kao, the man dubbed "Mr. Creativity" by *The Economist*, said it well: "What will be the collaboration space between human and machine while we figure each other's intelligence out? My Tesla today does a good job driving autonomously on the highway, but its preferences for where to put itself in the lane, when to bypass a car or take the foot off the throttle do not match my preferences and it doesn't really ask about them either."* As Kao suggests, we will have to adjust to the idea of giving up more control, and doing that requires a higher level of trust in the systems to which we delegate authority. If history is any guide, we will eventually have to grant more trust in the symphony of intelligences that Kevin Kelly describes, and the concert of instruments around us will play a sweeter and richer melody.

Yet our trust must run deeper than the AI systems that govern more of our daily details and decisions. If the scandals that enveloped Facebook after the 2016 US presidential elections and the revelation of Cambridge Analytica's inappropriate use of user data scraped from

---

* John Kao, "The Nature of Innovation Through AI," (lecture, HULT-Ashridge Executive MBA Learning Journey Silicon Valley, San Francisco, CA, Nov 2017).

the site tell us anything, it's that we need to be able to trust those who control the entire system. Who or what will monitor integrity, fairness, and equity in the back rooms and far off data centers we'll never see?

MACHINE JUDGMENT

Consider, for a moment, how much your life has changed from ten years ago. Today, you might scoff at the things that caused such stress back then. Perhaps a growing family has fundamentally changed your priorities. Hopefully, your expectations and experiences have become richer and more fulfilling. Regardless, they've changed. Now look ahead ten years. Can a world increasingly pervaded by thinking machines fully grasp your changing preferences and values? In the past, you might have set the preferences on your autonomous driving system to do whatever it can to dodge a dog that runs onto the road. Now, you have your spouse, two kids in the back seat, and elderly parents—a cargo of expectations and responsibilities riding along with you. Does the car know the dog no longer matters nearly as much as it once did? Compared with the responsibility you still feel for the dog, your responsibility for your kids and your partner carries far more weight. And while the AI system that powers the cars' navigation can track changes in your life, how does it crosswalk those behaviors into an accurate map of your values?

In every area of life, machines are making more decisions without our conscious involvement. Machines recognize our existing patterns and those of apparently similar people across the world. So, we receive news that shapes our opinions, outlooks, and actions based on inclinations we expressed in past actions, or the actions of others in our bubbles. While driving our cars, we share our behavioral patterns with automakers and insurance companies so we can take advantage of navigation and increasingly autonomous driving technologies, which promise to provide us new convenience and safer transportation. And as we continue to opt into more and more conveniences, we choose to trust the machines to "get us right." In many instances, the machine might get to know us in more honest ways than we know ourselves. But the machine might not readily account for cognitive disconnects between that which we purport to be and that which we actually are.

Reliant on real data from our real actions, the machine constrains us to what we have been, rather than what we wish we were or what we hope to become. So, what does the machine really know? Enough to really make an actual *judgment* about who we are and what we believe?

These days, even the humblest thermostat makes a sort of very simple judgment, regulating the temperature of the home even when it's devoid of its human inhabitants. Residents set the temperature range they consider comfortable, and then delegate the decision to turn the heater on and off. Yet, it's a completely physical machine. A spiral of metal, called the bimetallic strip, curls tighter when warm and loosens when cool, tilting a small bulb to the left or right depending on the temperature. If the bulb moves far enough, the bead of liquid mercury inside it connects two bits of metal, completing a circuit and kicking on the heater or air conditioner. More sophisticated thermostats have a clock and calendar integrated in them, regulating a home's HVAC system based on both temperature and time of day. Still, they're mechanical, so it's hard to think of them as making a judgment. Yet, in the simplest terms, that's what it's doing: making low-level decisions on one's behalf. We're delegating a decision, however routine or trivial, to a machine.

A smart thermostat moves a step up the chain. It processes the same factors as its predecessors but can amalgamate a wide array of additional factors—whether residents are active in the house, the weather, past heating and cooling patterns. It's not hard to imagine the Nest thermostat on the wall considering the spot-price of natural gas, the near-future effects of a preheating oven, or even the perishable groceries mistakenly left on the counter. From a technical standpoint, the idea of a smart thermostat making a decision for you remains a straightforward optimization process: weigh comfort against cost, or find ways to maximize one's well-being while minimizing, say, grocery spoilage. This doesn't require vast amounts of data and appears to involve rational judgments that we can understand, perhaps making it more palatable for us to say the smart thermostat makes judgments on our behalf. After all, if a homeowner sits on the couch all day and does nothing but optimize the balance between comfy temperatures and heating bill costs, he or she might make the same decisions the smart thermostat does.

But what happens when the machine starts incorporating factors beyond a person's sphere of awareness? Now it reports a home's autumn heating activity to the cloud, where the local natural gas company uses the data to more accurately predict demand for the winter. The regional geopolitical picture looks a little shaky, and the country that produces natural gas has threatened to put restrictions on its exports. The local gas company decides to reduce demand in autumn to build reserves for winter, and it opts to crank back everyone's comfort just a little bit—hardly noticeable to you, but valuable for the region as a whole. Now, the thermostat and the network of thinking machines to which it connects optimizes for an entire community, perhaps nudging the temperature of your home lower than you'd prefer.

Artificial intelligence drives the automated infrastructure of our lives. Increasingly, devices in our cars or our homes are connected to one another. They communicate with us, with each other, and with servers in the far reaches of the Internet. Over the next twenty years, those networked devices will interconnect with "smart" infrastructure, such as highways, airports and train stations, and city control centers for security and environmental protection. This Internet of Things, which started as innocuous machine-to-machine communication in factories, has steadily made its way into all kinds of public spaces. Virtually everything will have an IP address at some point, from your mattress to your trashcan to your shoes—all woven into one big network that fuses the physical and digital worlds.

AI-powered technology will make it all tick, ideally enabling more productive, richer, and safer lives. It will control the augmented reality (AR) glasses you will wear on a future vacation to Rome, showing you information and interactive graphics in the Colosseum and then recommending a nearby restaurant. Based on your spending patterns and a mood analysis, a future home-care robot will know whether you want (or deserve) a bottle of that expensive Chateau Lafite Rothschild wine that's been on your bucket list for a while, and have it delivered if you do. All these large and small decisions require the integration of personal data with supply chain information, infrastructure updates, and commercial availability. As such, these AI systems will exercise a certain economic power, itself loaded with inherent value judgments—some right, some wrong.

Companies already see the potential, of course, and many are rushing in with new technologies to better integrate all the disparate systems we use today. A firm called Brilliant has created a product that coordinates many of the smart systems in the home and makes them available in a more seamless fashion, integrating them into a panel it describes as "the world's smartest light switch." Most home systems in 2018 interface with humans through computer or smartphone apps, and the growing use of the Amazon Echo or Google Home has moved home automation to a new phase, says Aaron Emigh, Brilliant's cofounder and CEO. The next step that Emigh and his colleagues hope to drive is a smart system that's built in and native to the home. "It's a naturally evolutionary process with technology after technology, when it becomes part of what you expect for what constitutes a home," he says. "What we consider a home today is different from what our grandparents valued in theirs." Given that we interact more with light switches than almost everything else in our homes, they seemed like the most obvious place to integrate the new technology.

The Brilliant panel includes a screen on the front end, and a platform to combine and learn from many of the data streams produced within a home. Analyzing a breadth of data can help make the home more comfortable, more secure, more livable for older residents, and healthier as users have it monitor diet and sleep patterns. For example, Emigh says, AI-powered systems that bridge a range of data sources could provide stronger home security, at a lower cost, and with less false reports than current offerings. Audio-monitoring devices could flag the sound of glass breaking. Cameras could feed back into facial-recognition systems to distinguish between a family or a potential intruder. "You lose out on a lot of possibilities for the value you can provide if this data is not stored and accessible for use," Emigh says.

Yet, he readily acknowledges the concerns users might have about the security and privacy of all that data, especially coming out of a place as intimate as one's home. Physically, the smart switch has an opaque plastic panel that can cover the camera embedded in it. Digitally, Emigh and his colleagues realize it's impossible to completely lock down all the data going from multiple devices to multiple providers. To some extent, they're trying to make sure Brilliant doesn't provide yet one more attack possibility for hackers. But, unlike many companies, they set out to address security concerns at some of the earliest stages in their development—something many companies do only after trying to build and increase demand for their products. Different people will have different levels of trust, and Brilliant tries to accommodate that from the outset. "The change for me has been entirely positive—the convenience and comfort and enjoyment," he says. "But I agree with you, there are downstream consequences on everything we do. In some cases we can figure out what they are, but in others they're not foreseeable."

What Brilliant or any single company can't foresee is just how much these relatively simple devices, such as a smart thermostat, will reveal about the humans and environments they measure, nor how those individual factors will feed into broader environments and communities. Ideally that's what it's all about—a cognitive element in the house that reflects on your home life and helps enhance security, comfort, affordability, and entertainment as it integrates and manages these different resources.

That is still uncomfortable for many. Yet, we've already acceded to the idea of carrying around a far more powerful and complex monitoring device almost everywhere we go. A search engine on your mobile phone can tilt results based on your location, past and recent activity, and perceived goals. If you suddenly start to walk with a limp, a mobile phone's accelerometers and gyros could identify the change in your gait. And, if you happen to be in a high-risk group for falling, the phone can send out alerts, either encouraging you to stop or urging others to intervene before you let pride prevail and end up with a broken hip.

At some point along this continuum—from simple thermostat, to comprehensive and intelligent home-automation systems, to the phones we take everywhere—our regard for the system changes. Certainly, networks that integrate machine learning capabilities bring an increased level of intelligence and potential value, but at what point do they make the types of judgments that matter to us and, thus, require greater degrees of trust?

THE MACHINE WILL SEE YOU NOW

Self-awareness, image, and values help mold the complex mix of factors that shape our private and public identities. A rich and ever-changing blend of all our education, socialization, and past experiences, both at home and in the workplace, infuses itself into our view of the world and our unique role in it. Even the most introspective among us have little clarity of what's going on under the hood. Now, into this miasma of wonderfully mysterious identities, we introduce systems with a bizarre mix of their own—equally capable of identifying the subtle identity hints we can't or don't want to see and overlooking some of the most important facets of our humanity.

Isn't it hard enough for us to figure out who we are, who we want to be, and what we want the world to think of us? In today's hyper-connected and mediated world, it has become even harder to assess the identity of ourselves and others, and our natural biases and generalizations often spark frustration. Oona King, the global head of diversity at YouTube sees it in the push for diversity in programming. "It's not enough to feature more women or black people on TV shows," King says. "We have to get much more granular about this." The deeper push extends beyond the script into the way characters speak and how subtle behaviors and reactions are displayed. And here, AI-powered systems could help generate some long-overlooked humanizing insights. For example, to analyze how much bias is expressed in its programming, YouTube has started to use facial recognition algorithms to identify how much protagonists of a certain gender or ethnicity are represented in programming and how much or little they speak. "We have found that when a male is the main actor in a scene and has a female opposite him, the male speaks 90 percent of the time," King says. "But when a female is the main protagonist with a male present, the female only speaks 50 percent of the time."

Those sorts of frictions exist in so many of the factors and expe-riences that shape our lives and relationships. In our workplaces, human bias is notoriously pervasive in hiring decisions. Several new firms, including a Seattle-based company called Koru, offer platforms designed to help companies improve their interview processes and pro-vide better access to minority and other often-overlooked candidates. The traditional hiring process resembles an odd mix of investigatory interrogation and beauty pageant—one side trying to probe for flaws, the other trying desperately not to reveal any. Companies and job candidates can and often do switch roles, but the same sorts of narrow judgments, ill-formed opinions, and outright biases remain no matter who's trying to impress whom. So, Koru, HireView, and other compa-nies employ AI systems in hopes of facilitating better, more-objective matches between employers and candidates. Using facial and voice recognition software and AI algorithms to analyze a video feed of an interview, recruiters can pick up on subtle clues of comfort and discomfort, truthfulness, confidence levels, and overall appearance. They use this analysis to predict a candidate's performance against other candidates along a number of psychological parameters the employer deems key for the role in question. Then they compare the external candidates to internal employees who are already performing similar jobs well.

Koru amasses about 450 data points and, rather than feeding them into a single predictive model, runs those variables through a combination of five different models simultaneously, says Josh Jarrett, the company's cofounder and chief product officer. By testing multiple models on data for current employees, Koru can find the model that best identifies the factors that correlate with success at the company. It then uses that model to process candidates, zeroing in on what Jarrett calls "GRITCOP"—for grit, rigor, impact, teamwork, curiosity, ownership, and polish. In minutes, Koru can run the data and find the best model, then test candidates with a twenty-question survey and rank them on the probability that they'll be a good fit.

The system will not replace the human-to-human interview, but it can help counter individual and organizational bias, helping companies find employees from places they might not currently consider. For example, one of the AI models associates high performance with top colleges and universities. But that correlation might be an artifact of the managers' bias toward, say, Ivy League graduates. "The whole point of this was to widen the funnel," Jarrett says. "We'd get top Harvard candidates if we want them, so let's find people from other schools. Let's find the GRITCOP stuff." In that sense, Koru can help corporations escape the tyranny of the résumé or curriculum vitae, which emphasizes pedigree that creates structural barriers for equal opportunity and de-emphasizes important psychometric variables.*

Koru uses feedback from the models to help candidates, too. In surveys, job seekers regularly complain that résumés disappear into a black hole and they never receive feedback. So, Koru positions its

---

*   Terence Tse, Mark Esposito, and Olaf Groth, "Resumes Are Messing Up Hiring," *Harvard Business Review* (July 14, 2014).

services as a supplement to the interview process, returning advice specific to each individual and engaging them in an exchange that helps inform both Koru and the candidates who use the platform. A candidate whose job interview responses are analyzed by Koru or similar AI systems could learn from comparisons with other candidates, but also with successful existing employees, who are held to a higher standard of real-time behavioral and psychometric evaluation, rather than historical analyses of résumés that over-emphasize pedigree and reinforce social strata along education and socioeconomic status.

Meanwhile, the computer vision algorithms mentioned above assess minute clues in one's interview conduct that are often hidden to the human eye. After all, the human interviewer is also beholden to their own biases, notes David Swanson, former chief human resources officer of SAP North America and author of *The Data Driven Leader*, which looks at how to use data to drive measurable business outcomes. "We have found that all but the most experienced and trained interviewers often spend the first five minutes of an interview forming an opinion, and then spending the next 55 minutes reconfirming it," Swanson says, "rather than continuing their exploration of the candidate." Early evidence suggests AI systems such as Koru's can mitigate this and give better feedback to both interviewees and interviewers, but the call is still out whether that leads to demonstrable job success in the medium and long term, as the value and agility of a human employee in an evolving workplace environment becomes clear. The risk might be that corporations use data metalabels that measure narrow, near-term suitability for certain tasks but sacrifice the longer-term flexibility of an individual, and their fit for evolving corporate strategies or ventures.

Yet, the returns are promising enough to spawn a range of similar applications. In April 2018, Cisco acquired a start-up called Gong.io, which can record sales calls and provide and analysis to salespeople and their managers alike.* The platform allows the sales force to better identify key leads and how to improve their efforts to close on deals, including ways to view the best practices of top performers. Managers, on the other hand, can get a more transparent view of the field of

---

* Daniel Karp, "Delivering the Next Level of Sales Efficacy with A.I.," The Source (Cisco Investments blog), April 16, 2018.

potential deals and how their sales staff is working to close on them. Few salespeople anywhere report the raw, unbridled, and unpolished truth. After all, their ability to tell appealing narratives and believe in them is what makes them good salespeople. But it's critical that managers ascertain an accurate portrayal of the situation.

Relying in part on the judgment of Gong, Koru, and other AI platforms might give managers a clearer picture into the potential of candidates and existing employees, but the same systems might also displace some of the deeply useful human intuition that helps people work better and more closely together. An X-ray system of the mind can help avoid the imperfections in our human decision processes, alleviating certain biases in our cognition and consciousness, but they might also focus too narrowly on certain measurable dimensions. Discarding the richness of all the other, tough-to-quantify cognitive attributes—our creativity, inspiration, adaptability, and intuition—precludes our chance to consider the rich, emergent potential of our minds and the power of the constantly evolving human consciousness.

CONSCIOUSNESS, VALUES, AND THE ETHEREAL HUMAN SOUL

We all live inside our conscious minds, but few people spend as much time contemplating that curious and indefinable space as David Chalmers. A philosophy professor and codirector of the Center for Mind, Brain, and Consciousness at New York University, Chalmers treads in some tempestuous territory, yet he seems to exude an indefatigable calm. He assumes he'll never know a comprehensive answer to his field's primary question—what, precisely, is consciousness?—yet he's perfectly happy to keep striving for it nonetheless. His demeanor almost belies the intensity of thought, but one of the few conclusions he's reached with any conclusiveness: "Our understanding today will look awfully primitive 100 years from now."

Recently, though, he and other thinkers in the field of the mind and consciousness have moved away from a hierarchical sense of consciousness at the top of a ladder. Traditionally, the thinking might have gone from cognition, to intelligence, and finally up to whatever consciousness was. To the layperson, Chalmers describes it as the little movie going on inside your mind, the one that only you can see and hear. Yet, a rising movement among philosophers of the mind has started to look at consciousness as a more primitive phenomenon—essentially, as *any* subjective experience, rather than *the* subjective experience. Pain, then, is a basic form of consciousness, Chalmers explains.* Fish feel it. Infants feel it. Adult primates feel it. One recently developing argument in the field takes it even a step further. Integrated information theory, introduced by the neuroscientist Giulio Tononi, posits that consciousness is inherent in any physical system, deriving from cause-and-effect relationship between the system's constituent parts.

Whatever the true nature of consciousness and the continuum on which it might exist, Chalmers says, we might make some useful distinctions when thinking about the subjective nature of humans and machines. "Some systems have elements of cognition but aren't able to think about themselves, so maybe self-consciousness kicks in at the level of primates or something," he muses. "Then maybe that's another step in this chain." The chain might start at perception, or how we perceive the world. Chalmers suggests that consciousness comes in very early, with perception, but that many systems have conscious perception without the ability to think or reason. So, perhaps our next step is cognition, that ability to think and reason about the world around us. Then, we might move another notch beyond with self-cognitive reasoning, in which we think about ourselves and our cognition. And with that, we might sense the self-consciousness, subjective, "first-person" experience about ourselves.

Yet even these distinctions suggest a hierarchy that Chalmers might question. The debates and discussions that surface on YouTube or at various conference panels can veer off into some surreal hypotheticals and trains of thought, which might explain why Chalmers enjoys the field in all its enigmatic glory. But the same uncertainty raises a critical conundrum for the development of AI systems that will monitor us, make decisions for us, have opinions about us, and judge us. "The science of consciousness has really developed a lot in the last twenty to thirty years, and AI and computer science are in principle a

---

* Interview with the authors via video conference, February 20, 2018

part of that," Chalmers says, "but it's hard to approach consciousness directly because [AI developers] don't know what exactly they have to model. . . . In the human case, we can start with things we know are conscious, like other humans, and try to track back from there. But there's no piece of code we can write and then say, 'That's it. Now we have consciousness!'"

Rather, consciousness appears to be an emergent phenomenon, irreducible to its individual physical parts or a set of clear causes. The interplay of billions of neurons in our brains with the millions of sensory inputs they get through our bodies create thoughts of a higher order—almost like a space station of the brain, orbiting on a higher plane, clearly supported by the physics of earth's resources and atmosphere but hovering above. Human self-reflection might offer the best evidence of complex consciousness, even if we can't explain precisely how it happens or from whence it emerges, Chalmers says. But what's clear is that consciousness, like identity and the many other indefinable attributes that make us human, is not a fixed or finite variable. Even we, as humans, will try to chase a higher plane of increased consciousness through meditation, Tai Chi, spiritual engagement, experiences in nature, or experiments with drugs. Some of us merely want to think and argue better. Some wish to perform better in their professional roles and careers. Still others want to become more fulfilled in their relationships, seeking wisdom and growth toward a higher state of being.

All these pursuits require a heightened level of brain function, alertness, and awareness. One need not become metaphysical to find this challenge appealing. Keeping the brain fit and ensuring stronger mental health as we age is enough, thank you. It's not for nothing that brain workout start-ups like BrainGym and SharpBrains have gained big followings in the past decade. Of course, the insights that spring from these wells of digital pattern recognition and reflection are not always pleasant or ego-reinforcing. Like any good feedback, this can force people to think hard about who they are and what they want to become. And in this sense, especially, AI systems appear to approximate a more complex consciousness than we might otherwise grant a collection of silicon, metal, and code.

Even that anthropomorphic sentiment is enough to rankle Jerry Kaplan, who lectures about the social and economic impact of AI at

Stanford University and has written extensively about consciousness and artificial intelligence. Kaplan gives no quarter on the consciousness of machines: "It's a mistake to use that terminology," he says. "We don't know what human consciousness is, so to apply it to machines is completely inappropriate. And there's no evidence to date that it will ever be appropriate to talk about machine consciousness." It's not so much the concept that bothers Kaplan; it's the language. A computer program could model its own existence in the world and reflect on that simulation of its actions. A robot takes an action, fields the resulting inputs from the surrounding environment, and then sees if its action had the predicted effect. To that extent, there's a sort of metareflection occurring, but to ascribe to that an anthropomorphic "consciousness" gets Kaplan more animated. There's no higher plane that the robot's circuits create, no place where it can reflect critically about its existence, its place in the universe, or its feelings of satisfaction, doubt, or curiosity.

He accedes the idea of symbio-intelligence, but even there adds a limitation on how far one should go with the concept of intelligence in machines. Narrow AI is quite real, he says, and programs designed for specific tasks can easily and distantly exceed human abilities. "If you want to call that intelligence, fine, but it's not relatable to human intelligence that generates abstractions and representations," he says. "We wouldn't build these if they didn't exceed human capabilities, reduce costs and generally do things better. I'm not threatened by that. That's what we do. That's why we built airplanes and they're not like birds. They're 'smart' enough that once switched to autopilot they can keep themselves up in the air, much better than humans, but that doesn't mean they have consciousness."

Christof Koch might be willing to grant some basic level of consciousness in a machine, but he remains, as he quipped on a 2018 South by Southwest panel, "a biological chauvinist." Echoing Tononi, Koch, the president and chief scientist at the Allen Institute for Brain Science, suggests that consciousness is not an emergent property, but inherent in the brain and other systems. Take away the concept of the soul or the romantic exceptionalism of the human being, and the idea of consciousness in a machine or another organism might not sound quite so fantastical. But the depth of conscious experience varies widely, he argues. Current neuroscience research suggests that consciousness derives from fundamental causal relationships in the brain, and those relationships are exceptionally vast and complex—far more so than the comparatively simple binary relationships embedded in a silicon chip. And while higher levels of complexity might be simulated in a deep neural network, they can only be simulated. "You can't compute your way to consciousness," Koch said at the Austin, Texas, conference.

To produce something closer to human consciousness, he argues, one would need to create a technology with far more causal relationships embedded in its core architecture, perhaps some future successor to the current quantum computing technologies. "Once a computer is complex enough to begin to rival the human brain, then in principle, why should it not also have conscious experience?" he asks.* As his self-proclaimed biological chauvinism at SXSW attests, Koch believes we're far from any such technology. A vast gap remains between machines and the complex consciousness experienced by humans. "If it's not wet," he proclaimed at SXSW, "it's not conscious!"

The tongue-in-cheek line got a chorus of laughs from the SXSW crowd, including a chuckle from Chalmers, who joined Koch on the same panel. Chalmers is more circumspect about it all, of course, but he also stresses the distinction between articulating subjectivity and experiencing it. An advanced AI system might process different representations of the world and have the ability to communicate them, but that's not something it's feeling from the inside as a conscious system. And as researchers develop increasingly powerful and complex AI models, humans almost certainly will start to ascribe higher levels of consciousness to them, especially since their increasingly complex internal machinations make it harder for humans to understand what's happening within them. Intuition says the more sophisticated their behavior, the more likely we are to see them as conscious, Chalmers says. And as these machines ascend some perceived spectrum of consciousness—at least in the popular imagination—perhaps we begin to wonder what sorts of moral, legal, and ethical forbearance they might deserve.

So, why do these esoteric debates matter in our daily lives? As we turn over more decisions to machines, granting to them the power to make judgments about us or on our behalf, we must stop to remember that AI systems can approximate our experience but can't yet actually

---

* Kevin Berger, "Ingenious: Christof Koch," *Nautilus*, Nov. 6, 2014.

know the human experience. Empathy arises from consciousness and our ability to reflect on ourselves, to understand that others reflect upon themselves as well, and to construct a fundamental, mutual bond in that shared awareness. The completely mechanistic program can capture and reflect a theory of the mind, simulating emotion, identifying near-imperceptible physiological signs of frustration or satisfaction, and showing inexhaustible patience.

It might empirically understand where your mind is and meet you there, and it will do so without bias or emotional pushback, but it won't arrive with any true empathy for your condition in the context of the shared human condition.

THE SPACE BETWEEN MACHINE CONSCIOUSNESS AND THE HUMAN CONDITION

As professors, consultants, and directors of leading economic and technological initiatives, we regularly find ourselves behind a podium or on a dais. But in 2017, when asked to deliver a major broadcast presentation on AI and its influence on society, I (Olaf) felt a different sort of pressure. Coaches offered advice on content and optimal presentation techniques, but the greatest training and support came from my wife, Ann, an educational professional who knows how to reach an audience of laypeople. She had little experience with the topic, but she's intellectually curious, and that combination made her a great sounding board. More importantly, though, she also knew how to handle my mental state as I prepared. Early on, Ann pushed me to clarify my message and make it accessible, challenging me repeatedly on both key and mundane points in my presentation. "When you use the word 'machine,' it makes me think of motorcycles or washers, but not AI," she provoked, and rightly so.

Yet, the full depth of her support stemmed from her empathy, knowing when to switch from test audience to coach to cheerleader and becoming a motivational amplifier as the day of the performance approached. The beauty of the situation was that Ann could gauge when to switch from a critical to a supportive mode without me fully realizing that was what I needed. As an educator with training in literature and music, she could modulate between the hard data of the content flow and the softer behavioral aspects of preparing for a live stage performance. A spouse can be one's toughest critic and fiercest supporter, and Ann lived up to both.

It requires a certain depth of emotional intelligence and empathy—born of a complex human consciousness—to recognize when and how to modulate between those modes. For the foreseeable future, AI will not be able to do this. No system today exhibits a sense of when and how to switch between objective digital data and subjective information, such as the irrationality of emotions involved in performance, stepping to our side, accepting all facets of a tricky and often explosive mix, and then applying the subtle and smooth dance of the minds. They cannot discern the fine but critical line between the all-out preparation and the confidence necessary to produce at one's highest level, whether presenting a major business initiative to the C-suite executives or diving off the starting blocks for the 100-meter Olympic freestyle. The secret sauce of success includes a healthy dash of unwavering belief in one's ability to perform and to outperform others who might have the same level of aptitude.

Ann needed to provide the necessary critique to make my talk better, but she also needed the metareflection and the empathy to know when pumping me up would result in the better overall performance. Like most spouses, she could understand the tipping points of my frustration and weigh a range of contextual information, including the passing of my father a little over a year earlier. All these factors can be fed into a machine, weighted, and tweaked to provide improved outcomes, but the AI system can never share the dizzying sense of intellectual stimulation, opportunity, enthusiasm, and adrenaline that mixed together with my feelings of loss and stress to shape my emotional condition at the time. Yet, that was something another human, especially a close friend, knows instinctively and can mold into motivation.

Empathy amplified and converted the signals I sent to Ann, creating a shared experience and a productive, winning situation. What converts those signals in an AI system?

THE SPIRIT AND THE MACHINE

The concepts of spirituality and human exceptionalism don't go over so well in some AI circles. Many researchers don't buy the idea that humans embody something more than their constituent particles, which, they argue, could be rebuilt or re-created once we figure it all out. But in a world of 2.3 billion Christians, 1.8 billion Muslims, 1.1 billion Hindus, and hundreds of millions of other religious adherents,* any conversation about the similarities of and differences between humans and machines is incomplete without acknowledging the possibility a spiritual and unknowable something beyond our current understanding. While much of the technical conversation about AI steers away from religious or spiritual values, plenty of philosophers and theologians have taken a keen interest in the reemergence of AI and what it portends for the church, mosque, or temple.

In some cases, these values shape unique paths in AI development. In Japan, where traditional beliefs don't draw major distinctions between humans, animals, and other entities, the concept of extremely lifelike robots seems perfectly normal. The same sort of machine disturbs many people in Western society. Humans don't control the world or hold some higher plane of existence, explains Yasuo Kuniyoshi, director of the Intelligent Systems and Informatics Laboratory at the University of Tokyo. "We are part of many things—animals [are] just like us, or even non-animals, plants and stones and things like that," Kuniyoshi says. "It's just sort of an equal member of the world."

Christianity, Islam, Judaism, and many other religions hold human beings separate, as an exceptional part of a special creator-creation relationship. This concept then casts artificial intelligence in an intriguing theological light, viewing it as a new humanlike intelligence that people have created in their own image. Adding these highly capable AI systems alongside birds and dogs and humans doesn't diminish the value of any of those creatures, says Noreen Herzfeld, a theology and computer science professor at College of St. Benedict at St. John's University and author of *Technology and Religions: Remaining Human in a Co-created World*. The difference is that AI was created by humans in our image, so while "we believe we're passing along the

---

* Conrad Hackett and David McClendon, *Christians remain world's largest religious group, but they are declining in Europe*, Pew Research Center, April 5, 2017.

things we value, we're also passing along our faults," Herzfeld says. "We bear some responsibility for that."

The duty toward values, power, and trust that's intrinsic in a creator-creation relationship run through most of the Christian theological writing on AI, especially among those who, like Herzfeld, have backgrounds in theology and computer science. Russell Bjork studied electrical engineering at MIT before enrolling at the Gordon Conwell Theological Seminary back in the late 1970s. With a young family to support, he went over to see if he could get some work teaching at Gordon College, a Christian liberal arts and sciences college in nearby Wenham, Massachusetts. They invited him to teach one quarter of computer science and, as Bjork says today, that one quarter had turned into thirty-eight years by early 2018.

Bjork is hesitant to extend the Christian understanding of personhood to include smart machines, but he also takes issue with the idea of "locating personhood in a soul that's implanted in human being at some point between conception and birth, as if it's a separate creation distinct from the formation of the body," he says. Bjork sees personhood as something that emerges in the course of human development, so it's not inconceivable in his mind that a mechanical system could, in fact, obtain it. "I don't anticipate that in the near future," he says, "but it's not a theologically impossible idea." He recalls a time when the Artificial Intelligence Lab at MIT had a resident theologian, and notes that he shares concerns about the act of creation and the values, trust, and power relationships embedded within it. Will we create technologies that treat humans as valuable in and of themselves, or will AI systems discriminate against disabled, disadvantaged, or digitally disconnected people? "That which you value is what you embody in the things you produce," Bjork says.

Yet, a symbio-intelligent partnership between humans, nature, and now AI might produce far more than we realize from the outset, says the Reverend Dr. Christian Benek. Benek is a pastor, the CEO of The CoCreators Network, and a graduate of the world's first doctor of ministry program focused on theology and science, based at Pittsburgh Theological Seminary. He wonders whether AI systems might help people discover "the wheat in the weeds" that enhances their humanity. In the Bible, Jesus Christ uses the parable of the wheat and the weeds to explain how God finds the sacred among the profane.

Perhaps, Benek says, a complementary form of artificial intelligence might help us better understand what happens when a person senses the presence of God or something beyond their observable understanding. "You can look at this a lot of different ways, but why are we dismissing vast amounts of revelatory experiences that is potential data?" he asks. "We can't reproduce that data, but that information points to something beyond ourselves. Maybe with AI we can start to gather that data and put together information we haven't been able to quantify in some way. We might be just on front end of what it means to be human."

Benek's sense of discovery and possibility extends from his deep belief in a participatory, rather than a supremacy or escapist, form of theology. Participatory theology is based on a redemptive process in which anyone can participate, including humans and machines alike, he explains. In contrast, supremacy theology would provide no place for debate, critique, or new discovery, manifesting itself in developers who are unwilling to consider the ripple effects of their technologies. And escapist theology, he says, "has been demonstrated through some of the actions of Elon Musk and the late Stephen Hawking when they suggest humanity must flee to preserve its existence."

Whatever the possibilities for AI and human spirituality, the concept of symbio-intelligence echoes Benek's participatory theology and the spirit of discovery it embodies. We don't know much more about cognition and consciousness in AI than we do about cognition and consciousness in octopi or whales, but both animals clearly display intelligence and have things to teach us about our own values, trust, and power. What might we learn from artificial intelligence and the many pathways it takes around the world?

4

Frontiers of a Smarter World

Xianqiao Tong sleeps in Silicon Valley, but in his dreams he cruises the streets of Shenzhen.

Tong leads a young company called Roadstar.ai, one of the more intriguing, albeit lesser known, autonomous car start-ups in the United States. Founded in May 2017 by three engineers who previously conducted autonomous driving research at Google, Tesla, Apple, Nvidia, and Baidu, the team has set an ambitious plan to have a fleet of driverless taxis covering much of Shenzhen by 2020. They expect to have the first of their "robo-taxis" in service as early as the end of 2018, albeit with human backups behind the wheel, and then remove that person a couple years later and use a remote operations center to steer cars through situations they can't process autonomously, Tong says. They're already thinking about how to design the user experience for the remote drivers and the passengers alike.

Despite the fierce competition over autonomous vehicle technologies and the potential geopolitical sensitivities of a US company founded by residents of Chinese descent, Tong speaks openly about

Roadstar.ai's vision and prospects. He guards the recipe to the secret sauce, of course, but he's happy to explain how the start-up's platform fuses data at the sensor level, rather than soaking up all the data feeds and processing them on the back end, a process that reduces latency and allows more accurate identification of cars, bicycles, and other objects in the surrounding streetscape. He speaks freely because he and his colleagues, unlike their competitors in the field, enjoy the best of both worlds: access to the top tier of talent in the United States, and the government support and infrastructure development in China.

"You can get autonomous driving engineers in China," Tong says. "You can get them, but if you want the top ones you have to come here." He's convinced that will change over time, as Chinese academic institutions and private companies, backed by billions of dollars from national and regional governments, expand their expertise. But for now, Roadstar.ai can ride on the talent available in the high-tech capital of the world. That's hardly unique from an American or a Chinese perspective, of course. Waymo, Tesla, and a variety of traditional carmakers all have a significant presence in Northern California and fight tooth and nail for brilliant developers. Some of the battles for top personnel have spilled over into the courtroom.

Meanwhile, dozens of Chinese companies have set up shop in the United States, building new AI labs and recruiting from US universities and companies. Alibaba has kicked off a reported $15 billion investment in the DAMO Academy,* which will consist of seven labs in major high-tech hubs, including Hangzhou, Beijing, Singapore, Moscow, Tel Aviv, Silicon Valley, and the Seattle area. The collaborative effort, headed by Alibaba CTO Jeff Zhang, will include University of California Berkeley's RISE Lab, and other major US universities including MIT, Princeton, and Harvard.

What might set Roadstar.ai apart, however, is its connection back to China, where the founders recently launched a domestic company to develop and operate its robo-taxi services. "When the car becomes fully autonomous, the Chinese government has to have control of the technology or the car," Tong says. "So, it has to be someone with a Chinese background or citizen to work on this to make sense for China. That's the big advantage for us." And once it's there, Roadstar.ai will be able

---

* The DAMO Academy acronym stands for Discovery, Adventure, Momentum, and Outlook.

to work with a government that's both receptive and able to quickly create the conditions necessary for autonomous vehicles to work. For example, Hangzhou has started to build out what people are calling the "City Brain." Home to more than 9 million residents, the city government hopes to redesign its entire infrastructure in collaboration with Foxconn, Alibaba, and other high-tech giants. The City Brain will track residents via social media and surveillance technologies, but it also will begin to build out the infrastructure needed to support increasingly autonomous driving—including technologies such as Roadstar.ai's "man behind the curtain" remote driving centers.

These sorts of country-to-country differences can shape the direction of technical development of AI applications, but so too can the diverse values, notions of trust, and power relationships found in different parts of the globe. On the technical side, the lack of rapid, centralized infrastructure development means US companies tend to conceive of each autonomous vehicle as a separate entity, rather than a fleet backstopped by a central remote system. On the cultural side, we already have seen the limits of Americans' notions of values and trust emerge in response to fatal testing errors, most notably in March 2018, when one of Uber's autonomous cars killed a woman pushing a bike along the roadside in Phoenix. Most companies proclaimed a temporary moratorium on testing until developers and investigators could pinpoint the cause of the collision, and Tesla would later disclose that one of its Model X SUVs crashed into a concrete highway divider and killed a man while the autopilot systems were engaged, the second autopilot-implicated fatality for the company.[*]

The public outcries and potential litigation surrounding these sorts of incidents won't halt the groundbreaking technological leaps that Tesla, Waymo, Uber, Roadstar.ai, and other autonomous vehicle companies will make, but they could slow deployment and cast a far-reaching shadow across the industry worldwide. In the United States alone, cars kill an average of almost 100 people a day, but the idea of a driverless car killing a person seems far more troubling to most people, regardless of nationality. This deep skepticism puts heightened pressure on the companies developing advanced technologies

---

[*] Neal E. Boudette, "Fatal Tesla Crash Raises New Questions About Autopilot System," *New York Times*, March 31, 2018.

for self-driving cars, which must achieve far greater safety records than the imperfect human drivers they hope to eventually supplant. So, the responsibility of ensuring safety as a top priority is universal, requiring that developers make autonomy work under all sorts of road conditions, infrastructures, and regulations.

Yet, these companies cannot ignore the varied cultural norms and biases within the markets they serve. After all, these intelligent systems will make black and white decisions about many of the gray-area aspects of our personal and social systems. The fact that a Google image search for "cute baby" will yield nearly all Caucasian babies in the United States and nearly all Japanese babies in Japan raises enough serious questions about bias and discrimination. But what happens when those biases dictate something even more consequential than search results and aesthetics? How might those considerations affect Tong's robo-taxi service in Shenzhen or the future of Uber's autonomous vehicle testing in the United States?

THE FORCES THAT SHAPE THE WORLD'S DIVERGENT AI JOURNEYS

Building effective intelligent systems requires capturing the immediate objective of its users and beneficiaries, but also the broader society's values and norms into the system. A user must trust that the system will make a proper decision on its behalf. That's especially hard for complex or social tasks, such as evaluating employee performance, booking travel, shopping for nonperishables, or, at a more immediate level, engaging with a conversational assistant about these things. People are often unclear about their own objectives, and high-tech industries do not have a stellar record of understanding how those nuances can change from one moment to the next, let alone one culture to the next. The cultural and political sensibilities embedded in an AI system by developers in one country could power applications in every corner of the world, including in places where those sensibilities are impractical, offensive, or dangerous. That's because software is sneaky; it seeps into the nooks and crannies of our lives in often subtle, invisible ways.

We have a difficult enough time understanding how we can best deploy a new smart app on our phones, let alone how someone a mile away, a state away, or half a world away would grapple with the same technology. But we can boil down the overwhelming mix of influences into three essential forces that will shape the development of technologies and how they diverge from one place to another—the quality of data sets; the demographic, political, and economic needs of a country; and the diversity of cultural norms and values.

The quality and size of data sets and how they are used to train new applications will shape the power relationships between companies, governments, and individuals. Google faced a modest uproar when its Image searches for "beauty queens" returned photos of only white women. That biased data set offended people in the United States, largely because Google search results are so pervasive they can shape cultural norms and beliefs. But bias against minorities is not as big a concern in many other cultures, so, the old "garbage-in, garbage-out" adage remains in full effect. Biased, poorly constructed, or incomplete data sets lead to prejudiced or incomplete outputs, potentially excluding people from important civic, political, and economic dialogues, both within and between societies.

Countries often express their demographic, political, and economic needs in national strategies for artificial intelligence. As we note later in this chapter, these different national strategies, whether intentionally or not, sometimes align and sometimes diverge. In the United States and Israel, for example, the military drives much of the basic advanced-technology research even beyond defense applications. Yet, the United States has a far more prominent collection of Digital Barons, giant companies that dominate data and development. China has the same class of titans in Baidu, Alibaba, and Tencent, but the government plays a more active role in their strategic direction. Meanwhile, Japan and Canada take somewhat unique approaches to the development of AI-powered systems. Japan's deep background of Shinto belief lends itself to a broader cultural acceptance of alternative forms of life and consciousness, a fact that's often reinforced in its popular culture. Canada has created a more democratic and inclusive model of AI development thanks to the singular influence of a small group of developers and support from focused government science grants.

These many divergences stem in part from cultural notions about the balance of power and agency between individuals and institutions or communities. For example, when the global Institute of Electrical and Electronics Engineers (IEEE) released the first version of *Ethically Aligned Design*, a report about how to integrate values-driven, ethical thinking into every aspect of AI development, many Asian experts noted it was "extremely Western," says John C. Havens, executive director of the IEEE Council on Extended Intelligence. The second version, released in December 2017, pulled from a broader range of viewpoints and integrated ethical concepts from Confucian, Shinto, and Ubuntu philosophies. For example, an Ubuntu mindset focuses on forgiveness and reconciliation over vengeance. It recognizes, in the words of Nobel Laureate Archbishop Desmond Tutu that "my humanity is caught up, inextricably bound up, in what is yours." Considering those sorts of concepts, Havens says, "completely gets you out of Western thinking."

Widening the cultural and economic lens reveals an intriguing mix of AI-development models that have emerged around the globe. They often diverge, but they also overlap with one another, as most countries and regions share some elements even as they follow their own unique pathways. As we note below, the Cambrian Countries feature robust entrepreneurial ecosystems that are tightly linked with their strong academic institutions. The Castle Countries possess some of the world's most advanced scientific and technical minds, but they haven't yet developed the type of start-up environment that can build and scale the massive private-sector data titans. The Knights of the Cambrian Era have leveraged military spending and resources to build expertise across an array of AI uses, whether for defense or other purposes. The Improv Artists have found ways to encourage or develop unique applications of advanced technologies that address problems that many developing economies face. And then there are the outlier approaches seen in Japan and Canada, where developers have blazed their own unique trails.

In the years to come, elements of all these approaches will meld and conflict as researchers interact, companies work across borders, and nations press their ongoing quests for technological, economic, and political influence. We address more of the AI race in the next chapter, but to understand these models and how they might evolve in the future, we first need to understand just how data derives its power—and how massive troves of data have created new Digital Barons that exert an outsized influence on the development, regulation, and public opinion of artificial intelligence.

AS TRILLIONS OF DOLLARS FLOAT INVISIBLY BY

It's critical to understand how these cultural and political forces will direct the pathways AI development takes around the world (as we discuss below), but all those divergences fork off from the central notion that data is power. "Gigabytes? Terabytes? Bah, small potatoes," Cathy Newman writes in *National Geographic*.[*] "These days the world is full of exabytes—zettabytes, even." No one can quantify the precise amount of data generated in a day, but reasonable estimates are incredible. If one gigabyte is roughly akin to the information held on a ten-yard shelf of books, Newman says, the world filled about 2.5 billion of those shelves in the past twenty-four hours. No wonder *Harvard Business Review* named data scientist as the "sexiest job of the 21st century."[†]

That data comes in all sorts of shapes and sizes, from a Bangalore mother's online shopping preferences to a tsunami sensor's readings off the coast of Japan. For Internet companies, "life-pattern" data attracts the most attention. A 2015 McKinsey & Company report forecasts that data from all connected devices will create about $11 trillion in economic value by 2025.[‡] As part of the group contributing a significant portion of the $11 trillion, individuals might want a better understanding of that data and where it derives its value. Consider the lamp that might be shedding light on this page, and then imagine a translucent blue box around it. That box is a "semantic space," a description of what the lamp consists of, who built it when, and how they put it together. It might also describe the intended customer segment for that style or the suggested retail price. This semantic space appears when the factory produces the lamp, a digital representation of the lamp's

---

* Cathy Newman, "Decoding Jeff Jonas, Wizard of Big Data," *National Geographic*, May 6, 2014.

† Thomas H. Davenport and D.J. Patil, "Data Scientist: The Sexiest Job of the 21st Century," *Harvard Business Review*, October 2012.

‡ James Manyika, et al., "Unlocking the Potential of the Internet of Things," McKinsey & Company (June 2015).

characteristics that resides in the company's database. It lets management ensure better quality, encourage more sales, and develop an even better version at the next go round. The semantic space gets created when a factory makes the lamp.

Now, with the addition of more sensors, cheaper memory chips, and higher compute power, as well as new data processing capabilities embedded in the lamp itself, the semantic space tied to each individual lamp continues to evolve long after the sale. The lamp might connect with physical infrastructure, like electricity grids, building walls, and ceilings. It might connect to people in both passive and active ways, potentially monitoring usage patterns to optimize power consumption and convenience. It becomes one of a rapidly growing number of objects and entities that bridge between people and content, making someone a producer *and* a consumer of data. And that data has all kinds of value—social value for friends and family and the lamp company that wants to learn from that evolving semantic space. Combine that with the myriad other streams of data that we and our environments produce every day, and you begin to see the incredible economic value for employers, businesses, utilities, retailers, and employers. Suddenly, McKinsey's $11 trillion estimate, nearly the size of the entire Chinese economy in 2017, might seem conservative.

What seems like a laidback Friday night with a book at home is becoming, invisibly in the background, a tremendously complex integration effort for all the data streams that emerge as you switch on that cozy lamp, light the crackling fireplace, and turn on a little smooth jazz. Of course, companies will use that life-pattern data to try and sell you more of things you like, but they also will use that data to train and refine their AI systems to more accurately predict what you would most enjoy or benefit from at any particular moment. A home assistant might plan meals, suggest an outing for tomorrow, and remind you to connect with Tom, whom you haven't seen in a while. It might interpret the different ways you move around your house and analyze the intentions behind your patterns, perhaps warming the toilet seat when you awake on a cold winter morning. It might precisely calculate how your nutritional intake integrates with your exercise patterns, suggesting meals or snacks that optimize your fitness. We can't visualize all the ways these refinements and innovations will change our lives. Hopefully, companies use these systems to make better products, provide more convenience, and enhance people's productivity and leisure time, rather than just serve up more advertisements and enhance their profit margins. But having loads of data means having loads of power. What happens to trust in a society when all that data and power are concentrated in the hands of a few companies and governments?

THE DIGITAL BARONS

The Blob captured the imagination of moviegoers in 1958 and oozed its way into film culture in the decades since. In the movie, an alien lifeform travels to earth on a meteorite, hitting the ground somewhere in the woods of rural Pennsylvania. Two teenagers, Steve and Jane, played by the unforgettable Steve McQueen and Aneta Corsaut, see the meteorite crash over the hill and, driving over to investigate, almost hit an elderly man on the road. The anguished man had poked the meteorite with a stick, and now has the blob attached to his hand. Before the doctor can amputate his arm, the blob consumes the man . . . and then the nurse . . . and then the doctor himself. Freshly fed, the blob rolls on and over everyone who crosses its path, gorging itself until it becomes so large it threatens to consume entire buildings and the people within. But our heroes, Steve and Jane, eventually realize the blob doesn't care for the cold, and the town neutralizes it with fire extinguishers before the Air Force transports it to the Arctic.

To many of us, this is what the big Internet companies feel like—the giant data blob that rolls over every area of life, from retail to finance, from expert advice to dating, health services to car sharing. Nothing seems safe, it just keeps growing, and we have no fire extinguisher large enough to freeze it. Fortunately, we mostly reap the benefits of this particular blob, because it makes our lives easier, more convenient, and more interconnected. As these companies acquire ever-greater levels of detail about our lives the more we use their services, they make sure that their platforms feed our needs and keep us happy. The skeptic might say "addicted," as Salesforce CEO Marc Benioff does when describing social media users. "Intimate" might be a better term, because it embodies both the opportunities and risks inherent in deeper sets of life-pattern data. The collection and analysis of vast data sets can create greater intimacy between the Internet platforms and their users, as well as between individual users themselves. Google can provide more precise results for that search you were struggling to define. Facebook can help you find and reconnect with a long-lost friend. Amazon can suggest just the right product to supplement that gift for your spouse. Baidu, Alibaba, and Tencent do the same for Chinese users.

Of course, all that intimacy can produce negative results if companies don't live up to the data stewardship expected of them by users, societies, and governments. But even when they live up to both cultural and regulatory standards, how much data is enough data? Does that sort of intimacy start to feel uncomfortable if you know that different signals of your romantic evening are being tracked by your thermostat, lamp, surveillance cameras, and smartphone? Absent their closest friends, few people want to share that kind of intimacy with anyone, let alone the engineers at Google. And most importantly: How much power are we giving others by letting them follow our intimate patterns too closely, happily intermediated by that soft-glowing, energy-saving, house-monitoring lamp?

One might start to worry about how much data these Digital Barons have and how much they need to deliver the products and services we so readily consume. We already grant a remarkable amount of power in the data we share, and new AI models can amplify that power even more. Both generative adversarial networks (GANs) and one-shot learning systems increase the accuracy and precision of AI outputs. GANs pit two AI systems against each other, with one creating an artificial output—a fake image, for example—and the other comparing it with real examples in hopes of detecting the flaws. The competitive feedback loop between the two improves the accuracy of both systems collectively, like steel sharpening steel.

The one-shot learning model could vastly expand the breadth, increase the speed, and decrease the cost of machine learning. With this approach, a system already trained to recognize a variety of objects can start to identify similar things based on just one or a few examples, not unlike the way a toddler learns to avoid a hot tea kettle after touching a hot pan on the stove. A one-shot learning system trained to identify dozens of different weapons in a battlefield scenario could learn to recognize a different threat based on just a few instances of it, building on what it has learned about the characteristics of the others.

There's one caveat to these approaches, however. While both GANs and one-shot learning models enable deep networks to learn from few examples, both need the networks to be primed with human-labeled examples in the first place. Absent a more efficient and practical way to generate labeled data across a wide variety of new domains—and, frankly, even if such a source emerges—there's no reason for the Digital Barons to limit the flow of incoming data, and every incentive to expand it.

THE CAMBRIAN COUNTRIES

In recent years, Fei-Fei Li has become one of the AI world's most well-known academic researchers, with her papers cited thousands of times and her position as director of the Stanford Artificial Intelligence Lab. Li traveled a long way to get there, moving with her parents from Beijing to New Jersey when she was sixteen years old before enrolling at Princeton University to study physics. She went on to earn her PhD from Cal Tech and later led the development of ImageNet, a massive online database that contains millions of hand-labeled images.

That sort of data set seems almost routine today, but when Li first published ImageNet in 2009 it had emerged from a somewhat unusual notion—that a better algorithm couldn't make better decisions on its own; it also needed better data. With such a large, labeled data set now available to them, AI researchers started competing against one another in the annual ImageNet Challenge, where they could see whose algorithm correctly identified the most objects in the millions of images. In 2010, the winning system won with an accuracy rate of 72 percent (compared with a 95 percent average for humans). By 2012, though, an algorithm entered by Geoff Hinton and his colleagues at the University of Toronto unexpectedly knocked the mark up to 85 percent.* Their innovation? An interesting new technique called "deep learning," now one of the AI field's bedrock models.

These days, Li is still working to change the way the AI field thinks, not only about its technological challenges but about the people who solve them. She supplies one of the leading voices for efforts to increase diversity in the ranks of academics, developers, and researchers. As a computer science professor at Stanford and director the university's

* "From Not Working to Neural Networking." (special report) *The Economist*, June 25, 2016.

AI Lab, Li regularly mentors students whose gender or ethnicity is vastly underrepresented in a field dominated by Caucasian and Asian men. "I'm focused on diversity, not just gender or race, but diversity of thought," Li says. "I think we absolutely have to do it. I don't think this issue can wait. This technology is going to shape humanity and if we don't get the right representation of humanity to participate in this technology, we'll have negative consequences."*

Yet, for all the parallels between Li's professional standing and the broader AI ecosystem, none reflect as fundamental a force as her joint participation in academia and private-sector business, particularly in her role as chief scientist of Google Cloud AI. On sabbatical from Stanford during the spring of 2018, Li helped design products and services intended to extend AI capabilities to more businesses and individuals, with the idea of democratizing access to these advanced technologies. And much like her desire to facilitate greater diversity among the people involved in the field, she hopes to open AI development to people around the world. "Science knows no borders," she says, and she hopes to build a tighter collaboration between US and Chinese AI research. In fact, in late 2017, she helped launch Google's basic research lab in Beijing, her hometown.

Despite the cultural, governmental, and economic contrasts between China and the United States, the rich ecosystem of development and cross-fertilization within and between each nation have set them apart from all others. The tight ties between academia and private-sector serve as one of the most critical drivers of AI development in these "Cambrian Countries," with their unique combination of Digital Barons, leading academies, vibrant entrepreneurialism, and their cultural dynamism. Microsoft, Baidu, Facebook, Tencent, and other private sector companies have ponied up millions of dollars to attract the top academic talent in both nations, including Chinese companies landing students at US colleges and, to a lesser extent, vice versa. Those companies have put millions more into research initiatives at leading universities from Boston to Beijing. And while the same critical academic-business ties occur in other parts of the world, other regions haven't developed the same depth of interdependency that's found in China and the United States, which are well attached to each other's hip.

---

* Interview with the authors at Google's offices and via video conference, March 7, 2018

Shaoping Ma plays that bridging role at Tsinghua University, where he leads a joint research center with Sogou.com, the country's second-largest search engine behind Baidu. Ma focuses on search and information retrieval. In China, Ma says through an interpreter, pioneering research remains well behind the United States, but its companies have done a comparable job of developing a variety of applications, such as machine translations. The massive consumer base and the data it produces create more opportunities for Chinese start-ups, he says, and close cooperation with industry helps academic developers gain access to those large volumes of real data to fuel and advance their research. "I don't think we can overtake the US in five years in general," he says, "but in some application areas we could."

These academic-corporate relationships are vital to AI development because each side brings a different set of motivations to development. Jeannette Wing moved from her post in charge of Microsoft's global network of research labs to head up Columbia University's Data Sciences Institute. For her, the chance to tackle the major, long-term, basic research problems in AI sciences proved too tempting after years in a mix of industry, academic, and government positions. She still works closely with the other sectors, now with an eye on the benefits of AI, but she and her colleagues in academia don't have to worry about the same short-term, bottom-line pressures that companies face.

So, while industry has two big advantages over academia—"big data and big compute"—Wing can try to solve deeper fundamental questions about AI models and related issues. "Only the academic community has the luxury of time to understand the science underlying the techniques," she says. "And it's important to have this understanding because they [AI technologies] are already in self-driving cars and they are already being used in the criminal justice system, etc. So, the end users—drivers, pedestrians, judges, defendants—are going to be affected by this technology. It behooves the scientific community to provide a fundamental understanding of how these techniques work."

THE CASTLE COUNTRIES

Mikhail Burtsev wields the impressive scientific chops one might expect of Russian academia. He holds a PhD in computer science, which he earned while trying to model the evolution of human cognitive abilities. He focused on the theoretical side, trying to reconcile some Russian cybernetics ideas and develop a new way to get models of machine learning to interact with one another. "I'm still doing experiments with living neurons and what's happening with real neural networks [in the brain]," he says. "I have some theoretical part that stems from Russian neurophysiology, which is not really widely known in the West, and I'm thinking about how to incorporate those in the net architecture of AI agents."

Lately, though, Burtsev has shifted his expertise in new directions, creating a project called iPavlov, an initiative that simultaneously supports and belies some of the outside stereotypes about Russian artificial intelligence. On one hand, it's a deeply academic effort backed by government funding, having earned a grant in 2017 from the National Technology Initiative 2035. On the other, Burtsev is using that funding to create an open-source platform and database that will help developers build better conversational AI systems, something especially difficult given the nuances of the Russian language. In other words, one Russian friend told us, the government is financing an open-source project that will help everyone create better artificial intelligence.

"Russia has rather good potential in this field, but it's not realized yet," Burtsev says. "What we have is more or less good fundamental education and we have very good computer science skills for students, but on the other hand if you look at publication outcomes of AI, we'll see Russia is somewhere like fortieth place in the world and it's not very visible in the scientific landscape." The Russian government and President Vladimir Putin have recognized artificial intelligence as a vital piece of future geopolitical power, security, and influence, and it has started pumping more public funding into the field. Added to the pool of existing private investment, the public support has helped expand the resources available to start-ups, Burtsev says, but the country has not come close to its full potential given its academic standards.

The Castle Countries, which include Russia and Western Europe, have developed key expertise in certain facets of artificial intelligence, mainly on the academic side, hence the name "castle" in reference to the idiomatic "ivory tower." But none of them have the Digital Barons on the commercial side, which oversee the free flow of ideas, technologies, and capital from academia to the private sector, nor the dynamism to break down their castle walls. In places like the United States, Japan, and China, established technology transfer programs and a more robust entrepreneurial infrastructure help facilitate this. In Russia, however, conversations with several entrepreneurs reveal a start-up ecosystem that gets little support, with large, often state-backed companies and banks limiting the critical flow of capital to new ventures and the transfer of technology from academia to the private sector. Less tolerance for business risk, which is always inherent in early-stage tech start-ups, creates a chicken-egg problem for some small businesses in Russia, says George Fomitchev, the founder of Endurance Robotics, which develops a variety of laser-etching systems, chatbot software, and interactive robots. Investors want to see more established track records of sales, preferably to large Russian companies, Fomitchev says, but the big companies won't commit to products or services until they're fully baked. For one of their customer service chatbots, Endurance had to use the metrics it generated from work with British American Tobacco to pitch to Burger King, which needs a different set of products and strategies.

This leaves small start-ups like Endurance in a bind: Leave for other markets or keep trying to break into the established channels of capital, such as the oligarchies or the few accelerators and funding programs that exist. After Grigory Sapunov and his cofounder launched Intento, a platform that helps companies test different AI platforms and find the right fit for their operations, they realized they would need to break into other markets for sales and for the support their own business needed to grow. They have a marketable idea and have been building a client base. Using various public or proprietary data sets, they can test the many cloud-based AI platforms and find the ones best suited for a particular client. Then, their platform allows customers to use and switch between many of the providers, depending on the task, cost, or performance.

"The main disadvantage is that Russia is very far from the modern centers of the AI movement," Sapunov says. "A lot of Russian start-ups are trying to do something meaningful, but a lot of them are working for the local market, and it's hard for them to get into the global market." So, one of Intento's cofounders set up shop at an accelerator in Berkeley, California, looking to tap into the US market and the pool of resources in the Bay Area. Too few people run their own businesses in Russia, Sapunov says, and the California base gives them a connection to the entrepreneurial spirit there. Back home, he tries to poach talent from Yandex and some of the other large high-tech companies in Russia. "There are lots of interesting jobs in start-ups, but people want to have a safe place," he says. "It's a problem because AI is a field that's very dynamic. To catch the wave, you have to take some risk in your activities."

Western Europe exhibits some characteristics similar to Russia, including a somewhat wider gap between academia and industry than that in the United States and China. However, the European model—which combines a high degree of scientific and academic competence with its sizable manufacturing base and more open approach to data sharing—differs from Russia's in two significant ways. First, Russia tends to channel more of its computer science and mathematics prowess into defense and national intelligence than do the EU member countries. Second, and perhaps more significant, Europe has developed a much more robust entrepreneurial ecosystem than their Russian counterparts. While its start-up environment trails that of the United States and China, Europe has seen the emergence of several key digital hubs in Berlin, Hamburg, London, and Paris. Tallinn, the capital of Estonia, has become a cybersecurity hub and one of the world's most advanced centers of digital government. Gothenburg, Sweden, and Helsinki, Finland, have launched Nordic AI initiatives that facilitate open collaboration across academic and entrepreneurial sectors. Even some off-the-beaten-path places feature small but important pockets of innovation, including Lugano, Switzerland, a picturesque town of 60,000 people that's home to AI pioneer Jürgen Schmidhuber and the Dalle Molle Institute for Artificial Intelligence Research.

Still, Europe remains a long shot away from translating these pockets into formative and global economic powerhouses on the scale of the United States and China. Much of its current AI development tends to flow through manufacturing and other traditional industries. Add the European Union regulations on data privacy and security issues, and the innovative environment gets a bit more restricted (regardless of what one believes about the merits of those data-privacy rules). Damian Borth would like to see a more nuanced set of regulations, one that would keep protections where needed without stifling industry and start-up activity. Borth, director of the Deep Learning Competence Center at the German Research Center for Artificial

Intelligence (DFKI), suggests four classifications. If an AI system has an impact on human life, it slots into the A category and is regulated more closely. If it can impact the environment but not directly threaten people, it slots into Class B, and so on toward less restriction for systems that can do less harm. "If you want to have same market size as the US, Borth says, "you have to go to all of Europe, and then you have to deal with all the regulatory systems." So, for all the work done on artificial intelligence across the region, entrepreneurs still slip away to bigger markets.

What Germany and most other European countries do well is "the boring stuff," Borth jokes, like manufacturing. The game for personal data is lost. Starting a European version of Google, Amazon, or a continent-originated social network would certainly be an uphill battle (albeit one that the privacy-conscious part of the world might welcome) in a field in which "winner takes all" and the installed base of consumers plays a decisive role. So, AI development there tends to build around customer transaction data, data gleaned from the Internet of Things, and business-to-business applications.

Still, Europe's diversity of institutions and funding sources, as well as an emerging EU effort to develop a digital commons for technological innovation, makes it an interesting alternative model. More institutions, companies, and governments across the continent are helping in that evolution. Borth and his colleagues at DFKI, for example, work closely with industry to get better access to data and help companies integrate new AI models into their processes, from business operations to human-friendly robotics in the factory. England has become home to critical outposts for some of the world's Digital Barons, including DeepMind, which Google acquired for about $625 million (£400 million).* And while it lacks homegrown Digital Barons and the sheer volume of entrepreneurial activity in the United States and China, London and some of England's top universities have become rich seedbeds for entrepreneurial spinoffs, in many ways similar to the close academic-commercial relationships in Silicon Valley. From 2012 through the first half of 2016, the United States invested $18.2 billion in

---

* Samuel Gibbs, "Google buys UK artificial intelligence startup Deepmind for £400m," The Guardian, Jan. 27, 2014.

AI, China $2.6 billion, and the U.K. $850 million, according to a Goldman Sachs report.*

The very roots of AI reach back to Alan Turing and Bletchley Park, and universities such as Cambridge and Oxford already have established some of the world's most prominent high-tech research centers that host renowned experts such as Nick Bostrom, Jaan Tallinn, and Huw Price. Concerns about AI's effects on Britain's economy and its people have permeated the government, as well. In spring 2018, the British government joined with private companies and investment firms to commit $1.4 billion toward AI development. The money will support several initiatives, including UK-focused investment funds, an AI supercomputer at the University of Cambridge, and a new Center for Data Ethics.

The ethics center, for example, will help address some of the stickier problems of AI development, including what Lord David Puttnam calls "society's retreat from complexity." "The net effect of reliance on artificial intelligence could find us looking for oversimplified answers and solutions to complex problems," says Puttnam, a member of the House of Lords and the Oscar-winning producer of *Chariots of Fire*. Having dedicated his career to the exploration of societal issues in film and TV, he is part of the committee of the House of Lords that crafted the recent report on the UK's readiness for AI.† His key takeaway from the exercise was that our discourse increasingly shirks complexity, he says, looking for quick, simple data-endorsed fixes instead. "But many decisions in human life require longer reflection and awareness-building and should be deliberated and debated on an ongoing basis. Not all of those debates can or should be resolved quickly." Thinking machines can return clear-cut answers, but often without drawing attention to the many issues where "there isn't a winner, but a dialog and a synthesis and a negotiation," he says.

Ultimately, all these initiatives are designed to keep the United Kingdom at the forefront of the AI field, an effort that becomes increasingly important as the country finalizes Brexit and leaves the European

---

* Piyush Mubayi, et al., "China's Rise in Artificial Intelligence," Goldman Sachs (August 31, 2017).

† *AI in the UK: Ready, willing and able?* House of Lords Select Committee on Artificial Intelligence, Report of Session 2017–19.

Union, Puttnam says. If it leaves, the United Kingdom would limit its access to the largescale market and data availability that drives most commercial AI advancements. For the moment, the British government has passed legislation in spring 2018 that adopts much of the EU's data protection framework and expands the powers of its Information Commission to enforce those provisions. "In terms of our desire to strike a careful balance between economic growth and ethical safeguards we are in fact currently well aligned with Germany, France and Canada," he says. "Together, we carry much more weight in the emerging world of thinking machines. But what weight will the UK's voice carry in the global governance of it once we leave the EU?"

French government officials have pushed forward assertively on country-level programs, especially after the election of President Emmanuel Macron. In the spring of 2018, Macron said the government alone would pledge $1.85 billion over five years to support AI research, start-ups, and the collection of shared data sets. Macron told *Wired* magazine that the clear leaders in the field lean two different directions—the United States toward the private sector, China toward government principles—so France and Europe have an opportunity to find a middle ground.* The French are a notoriously techno-critical society, but Macron hopes to create an interdisciplinary effort that would provide a new perspective on which AI is built. "If we want to defend our way to deal with privacy, our collective preference for individual freedom versus technological progress, integrity of human beings and human DNA, if you want to manage your own choice of society, your choice of civilization, you have to be able to be an acting part of this AI revolution," he says in the Q&A. "That's the condition of having a say in designing and defining the rules of AI. That is one of the main reasons why I want to be part of this revolution and even to be one of its leaders. I want to frame the discussion at a global scale."

In this sense, Macron has forged ahead of his counterparts in continental Europe, asserting a national strategy on AI that includes government leadership on behalf of the individual citizen and aspirations of making Paris the primary research hub for AI development

---

* Nicholas Thompson, "Emmanuel Macron Talks to WIRED About France's AI Strategy," *Wired*, March 31, 2018.

in Europe. The German government has taken a more conservative, bottoms-up approach, seeking first to solicit the views of its industrial and scientific establishments through an AI summit at the Chancellery, the power hub around Chancellor Angela Merkel. The government now includes two high-ranking officials to lead the charge on digitization—State Minister Dorothee Bär and a department head, Eva Christiansen—along with a handful of digitization initiatives across its science, economic, and labor ministries. Meanwhile, these and most other Western European governments have noted the critical need for cooperation, even beyond EU initiatives, and several have started discussing potential partnerships. Some of those burgeoning efforts, such as a new French-German Center for AI, have been delayed because of other pressing matters, such as Brexit and immigration. But across the board, government officials in both France and Germany tend to believe the humanity of each individual citizen is an indispensable component of a vibrant democracy. So, in their view, the global AI competition becomes a race to simultaneously preserve the pre-eminence of human intelligence and European sovereignty.

THE KNIGHTS OF THE COGNITIVE ERA

The Israeli Defense Force knows it can't keep every nefarious actor out of the country. So, they make it as hard as possible to sneak through, and then deploy a sophisticated internal net to quickly identify and catch as many threats as they can. Yossi Naar took the same mindset from his work for the Israeli military and applied it to a cybersecurity start-up. Rather than build a higher digital wall—something plenty of other firms already try to do—Cybereason analyzes everything attackers might do after they find their way in, identifying them and rooting them out of the environment. That requires both advanced technology and a deeper knowledge of how hackers operate, the cofounders say. "There was this old point of view, where you could clearly define right and wrong, and then figure out how you build a higher wall," Naar says. "But in our nation-state background, we've known a lot of things to be simple and true: a. you can always get in; and b. the biggest and most difficult question for the attacker is what they do after they get in."

As Naar attests, the Israeli military has a deep influence on the development of a wide range of AI applications in the country. Every citizen must join the military, which developed an extremely effective system to identify top talent and track them into its sophisticated high-tech training programs. So, the research and development conducted there serves as a de facto training center and incubator for AI talent, he explains. Many of the country's leading high-tech start-ups were launched by partners who met while serving their country. It's an intense educational program, a full-time job in the classroom, and then graduates go to work on some of the most advanced technology platforms in the world. Through the reserve forces, others with extensive AI or digital experience cycle back to mentor in a "reinforcing system that brings knowledge in and takes knowledge out," Naar says. "That gives smart young kids a lot of resources to work with, which you don't get as a 21-year-old in college."

The defensive military perspective, especially in a country such as Israel, naturally generates nonmilitary companies that think in terms of defense, such as Cybereason. Yet, it also spawns an array of different ideas and talents born from its education and research. Of the 2,500 or so start-ups in the country, Naar estimates, about 500 to 700 are security related. For example, the intelligence community and its focus on informational analysis has helped power a great deal of research into big data of all types. And then there's the array of AI-powered health-care applications emerging from Israeli entrepreneurs as well.

While the United States doesn't have the same breadth of high-tech education within the military, it has developed its powerful links between national defense and advanced research through the Defense Advanced Research Projects Agency. The United States, Israel, and China stand out among the Knights of the Cognitive Era—countries in which defense-based innovation radiates into academic and private spheres, driving a range of peaceful and commercial applications. The United States doesn't conceive of every citizen as a soldier like China and Israel do, but it maintains an inextricable link between the military and civilian sectors. The US defense agencies shop in Silicon Valley, but they don't expect Silicon Valley to carry out their battles. Likewise, DARPA remains one of the world's premier facilitators of cutting-edge high-tech research, funding researchers who push the state-of-the-art on everything from autonomous vehicles to neural microchip implants and sophisticated systems analysis (e.g., climate change) to cybersecurity.

After her tenure as a program manager at the agency, Kathleen Fisher, head of the computer science department at Tufts University, observed DARPA's 2016 Cyber Grand Challenge, an open-competition tournament in Las Vegas. The competition pitted teams against one another, with each trying to defend a set of programs on their own systems while hacking the programs running on the other's. So, one team might write code to automatically patch their programs, and then figure out how to exploit those findings against their competitors. But in one intriguing twist on this capture the flag type of scenario, seven teams participated in a play-in tournament designed solely for fully automated systems. They had to design programs that could automatically protect and attack systems without human intervention during the game. One team's system "found a vulnerability, found the patch, patched itself, and launched the exploit against another team," Fisher says. "And while that was happening, yet another system identified that attack, reversed engineered a patch from that intercept, and patched itself. That all happened within 20 minutes."

The winning AI team out of Carnegie Mellon University competed the next day against the human teams, and it started out well because it could work faster. Over the course of the whole tournament, though, it fell to last place because humans could generalize and process a variety of different hacking concepts and strategies. "This will change," Fisher says. "Computers will beat everyone eventually. People are still better at exploiting software than computers are right now."

Like the Israeli Defense Force, though, DARPA's programs stretch well beyond cybersecurity and digital attacks. In fact, one of its key AI-related initiatives hopes to crack a problem plaguing just about anyone working in the field: developing an AI system that can explain how and why it comes to its decisions. The concept of *explainable AI* has baffled experts as these systems have become more complex. While thinking machines can learn on the fly and process massive and complex sets of data, developers still don't know exactly why the machine decides that one picture depicts a wolf and the other a husky. In one infamous example, researchers tried to infer an image-recognition system's reasoning by tweaking the input and seeing how it affected the output. They discovered that the neural network identified some huskies as wolves because they were sitting in snow.

While explainable AI has definite defense implications, DARPA's funding of work in the field has a broad range of ripple effects on how AI systems interact with humans, says Wade Shen, program manager of DARPA's Information Innovation Office. Plenty of machines can generate accurate decisions, but they're not put into use because people can't trust them. "Explainability" is plausible for certain types of AI models, but we're not close to understanding newer, increasingly complex technologies, Shen explains. So, while humans do quite well understanding cause and effect models, they're far more limited when those relationships depend on a massive number of variables, as they do in climate models, for example. "Machines might be able to build models of very complex processes to take into account thousands of variables and make decisions that humans just can't comprehend cognitively," Shen says.

Ultimately, we might need machines who understand and can interpret other machines for us in ways that simplify their inner workings for humans. Even many of the most elite and well-trained human minds struggle to understand how or why an AI system predicted stock prices to rise or fall. We still put our faith in many of these applications. But as these systems gain an ever-more pervasive role in our lives, we'll have to ask whether we want modeling capabilities we can never understand or predict, and how much control we're willing to give them. If self-consciousness is a higher form of consciousness because it reflects upon itself, as the philosopher David Chalmers suggests, machine consciousness is the equivalent of a toddler we're proposing to task with, say, genetic engineering or other analyses of monumental consequence.

THE IMPROV ARTISTS

AI-powered object recognition has become a popular application for e-commerce and related companies around the world. Take a photo and then click on the object of interest, and the app identifies the product and lets you know how or where to buy it. The big Digital Barons could do this for years, but newer companies in developing markets are taking it in new directions. Like Grabango and others in the United States, the Chinese start-up Malong has put it to use in the supply chain to help track and inspect shipments. They envision a time when a shopper could push a full cart of groceries out of the store, and its systems would identify all the items and automatically charge the customer as he or she walks out the door.

In Nigeria, Gabriel Eze is hoping a similar machine learning application can help open the web to fellow citizens who can't read or write in English, or at all. He and his colleagues at Touchabl currently focus on e-commerce sales—someone sees a purse they like and clicks on a photo to find out what brand it is and where to buy it. They make money by getting retailers and brands to pay for placements. "Maybe you have a broken part in your car, but you don't know what it is," he explains. "You can use Touchabl to find out what it is." If Touchabl hasn't already labeled it, it will search the web for a comparable image—a clear step beyond a random search. Eze also hopes developers will build on the platform, with designs to help informal merchants offer wares online via image or, if combined with language processing, help blind or illiterate residents access online information about the objects and environments around them.

He even imagines a time when similar systems could use photos people upload with their smartphones to diagnose health problems, such as cataracts. It turns out that's exactly what CekMata is going in rural parts of Indonesia. The archipelago suffers from high rates of blindness due to cataracts, with an average of one person losing their sight a day, says CMO Ivan Sinarso, citing World Health Organization statistics. Doctors can diagnose and treat cataracts long before blindness sets in, but cataract patients in rural parts of Indonesia rarely seek medical intervention, thinking it too expensive or ineffective. So, CekMata targets the younger generations of Indonesians, many of whom carry smartphones, enlisting them to take photos of their parents and grandparents and upload them via an app or website.

CekMata's systems can identify likely cataracts, and then recommend doctors who can confirm the diagnosis and prescribe a course of action. (Clinics pay for placement on the list of recommendations, Sinarso says.) In its first eight months online, the company helped about 100 rural patients identify and treat their cataracts, but the system can scale up to serve as many people as can upload selfies of their eyes, Sinarso says. And as it expands, the company will be able to track patterns, finding areas where people display problems at a higher incidence rate and alerting health authorities who could intervene.

Curtis and Mechelle Gittens also hope to address a critical health issue in their island country, but they're developing an entirely new AI model to do so—something that caught the eye of the IBM Watson AI XPRIZE judges, who advanced their team, called Driven, into the second round of the competition. The husband and wife duo are creating "psychologically-realistic virtual agents" to help model the thought patterns and behaviors of diabetics in Barbados. "By questioning the agent, you could actually identify an extroversion or introversion personality trait, for example," explains Curtis Gittens. "So, by simulating things like emotion and emotionally driven responses to stimuli, you'll be able to take this psychologically-realistic agent and almost query it as if you're a psychologist, and it would present traits as a human would."

Driven would take the patient information it derives from a survey of personality traits and behaviors, and then encode that to create a psychological representation of that person in a sort of virtual mind. Clinicians can run various what-if scenarios on the virtual patient to help identify ways to nudge real patients and keep them on their course of treatment. "We believe we'll be able to identify 'trigger' memories that are the root causes of behavior, so a doctor can work on the real factors that affect behavioral change," Curtis says.

Hope often springs from unlikely places, as illustrated by the popular William Gibson quote: "The future is already here, it's just not very evenly distributed." These Improv Artists of artificial intelligence—countries such as Indonesia, Nigeria, Barbados, and especially India—are developing new AI technologies or, more often, leveraging models to solve longstanding health, infrastructure, and other problems common in developing countries. Most of the world's largest high-tech players already have significant operations in India and see it as a massive digital opportunity. According to a report by Capgemini, 58 percent of the companies using AI technologies in India have installed them across a wide range of operations, putting it in third place behind United States and China in terms of the scale of deployment.* The country has seen a bloom in health care applications that integrate machine learning

---

* Bhaskar Chakravorti, "Growth in the machine," *The Indian Express*, June 20, 2018.

and other AI models, in many cases to address some of the most basic barriers to medical care.

India is home to more than a quarter of the 10.5 million people who suffer from tuberculosis, says Prashant Warier, the CEO of a medical imaging start-up called Qure.ai (pronounced "cure"). Many of those cases go undiagnosed, and even more of them get diagnosed very late, leading to the further spread of the disease. The problem, he says, is one of time. Rural patients will suffer for weeks with symptoms before traveling hours to a clinic to get tested. The doctors will order chest X-rays to search for signs of the disease, but because of the scarcity of radiologists, it might take a couple days before the physician gets back the radiologist's read of the scan. By then, the patients have returned home, making it difficult to contact them and perform a microbiological test to confirm the presence of TB.

Qure.ai helps compress that diagnostic process into a single day, Warier explains. Its platform can scan an X-ray image for abnormalities and return results in seconds. If the image reveals signs of tuberculosis, then the doctor can order the appropriate but costly microbiology tests to make the conclusive diagnosis. The whole process gets compressed down to a matter of a few hours, and sick patients can go home with the medicines they need to cure what ails them. Qure.ai focuses specifically on chest X-rays and head CT scans, automating the image analysis, helping identify abnormal pathologies, and then prioritizing critical cases for radiologists to review on an expedited basis.

They focus on those two core imaging procedures, but the platform is opening up an array of other possibilities, Warier says, including a move to other markets, such as the United States, and audits of a hospital's radiological diagnoses. "We can process all the X-rays a hospital does in a year, and do so in a few hours," he says. "And then we can compare that with the diagnoses on the written report using natural language processing. So, we can immediately compare and tell the hospital that there are these, say, 100 X-rays that were incorrectly reported."

Warier already sees a rapidly growing community of AI developers in India who will take these technologies in new directions. While still definitely behind China and the United States, the widespread and open availability of AI models and computer technologies makes the development of a vibrant industry a matter of talent and data. Qure.ai worked so well because it could access about 5 million relevant medical images and then tweak them, rotating them or cropping them in different ways to expand its training set even further. "We have done a lot of cutting-edge research on model architectures because our problems are different," Warier says. "Interpreting radiology images is a much more challenging task than decoding an image of a cat or a dog." Almost all the current AI research focuses on images that are about 100 times smaller than a typical chest X-ray, he explains. However, several open source repositories exist, and almost everyone in the field wants to publish their work, so there's a lot of open literature. Warier and many others will continue to build and innovate on top of what's already available. "For India, people here can work on the bleeding edge with the latest technologies and be up and running in no time," he says. "There's a lot of democratization and a huge amount of talent that can take advantage of this opportunity."

Not unlike Qure.ai, the talented developers at SigTuple also used AI systems to tackle a major time gap in rural health care. The company's Manthana AI engine can digitize and analyze certain pathology tests, returning reports in minutes, passing them along to pathologists to review and then on to doctors to make treatment decisions in far less time. That speed can have profound effects on rural care for acute cases of Dengue fever or other diseases. More than 100 million Indians live more than 100 kilometers from a hospital, with access to bare-bones clinics that don't have diagnostic equipment. Up to four or five hours can elapse while an ambulance is dispatched, arrives, and transports the patient back to a hospital. SigTuple aims to put its equipment in the clinics, allowing for a rapid enough diagnosis that ambulances can be dispatched with medicines to help treat the patient while on the return trip, essentially halving the time to care.

As in Nigeria or Barbados, the country still faces enough basic infrastructural problems that could hinder the application of AI solutions. In many cases, rural clinics don't have the connectivity or proper training to make use of technologies. While mobile connectivity is widely available, broadband speeds are not. The Indian national government doubled its budget for AI, 3-D printing, and other advanced technologies to $477 million in 2017,* and its national digital identifi-

---

*   Ananya Bhattacharya, "India hopes to become an AI powerhouse by copying China's model," *Quartz*, Feb. 13, 2018.

cation system, called Aadhaar, has opened possibilities for a range of data-driven financial and health care services that require online verification of individuals. But Indian states have significant control over the application of policy on the ground, so top-down initiatives don't fly as well as they do in, say, China. And, despite rapid educational and economic gains across much of Indian society over the past fifty years, about 300 million of its 1.2 billion residents still live in poverty, says Srikanth Nadhamuni, CEO of Khosla Labs in Bangalore. Any use of AI or other advanced technologies needs to focus on "the bottom of the pyramid where there are significant challenges and compelling needs," Nadhamuni says. "Making quality health care affordable to the rural poor through AI-enabled diagnostics with smartphone-based sensors and delivered at *kirana* (mom-and-pop stores) could transform the country's health care."

The same rural and impoverished chasms that technology has started to bridge remain barriers for sustainability. Almost two-thirds of the Indian population lives in rural areas, but 70 percent of the country's GDP churns out of its mega-cities, including Mumbai, New Delhi, Bangalore, and Hyderabad. AI-powered systems can overcome some of these barriers, but how can the companies developing the platforms make money from a population that has so little? And since these places likely won't be primary markets, how will companies adapt and adjust their applications to accommodate different cultural contexts, such as the dozens of different dialects spoken across the country? But the question isn't whether rural or poor Indians will accept artificial intelligence, Nadhamuni says: "People will take to solutions that solve their problems."

The concern with any technology, though, is its potential to create as many problems as it does solutions. New AI systems can democratize access to resources and amplify voices across class and income boundaries, but their benefits won't accrue equitably to everyone they touch. Rural Kenya's smallholder farmers have little to their names other than a tiny plot of land, a few tools, and a mobile phone to keep tabs on market prices, weather reports, and crop conditions. That digital access has given them far more control over their livelihoods. However, it also extends the widening advantage of the global technology elite who supply it. The digital intelligentsia that creates and sells technology accumulates disproportionate power, and that accumulation will only accelerate as the capabilities of these products and platforms increase, as is the case with AI.

As companies analyze data on global weather patterns, agricultural production, market prices, and infrastructure conditions, they can quickly shift global resources from one market to another. Few rural farmers and other digitally disadvantaged populations fully understand the global machinations that investment banks, global food conglomerates, and high-tech firms will play. So, while the poor Kenyan farmer or the rural Indian mother of four might benefit from the technology in their hands, they have far less opportunity to maximize their power or give voice to concerns about the ethics of food justice or global income distribution.

ASTRO BOY

Astro Boy made his manga debut back in 1952, but he lived in a future, science-fiction world where humans and robots coexisted in harmony. An android with human emotions, Astro was created by the head of the Ministry of Science to replace his lost son, who died in a self-driving car accident. The protagonist, known as "Mighty Atom" to the many Japanese who followed the series over the next sixteen years in *Weekly Shōnen Magazine*, would soon disappoint the inimitable Dr. Umataro Tenma, who realized the android could not replace the void of his lost boy. Astro would be sold to a robot circus, saved by a magnanimous professor, become part of a robotic family, and set off on various adventures.

The 112 chapters of the series, along with the subsequent remakes and spin-offs it inspired, now rank among the most influential forces in the history of Japanese manga and anime. But *Astro Boy* also provides one of the earliest pop-culture references for the more-symbiotic mindset that Japanese citizens hold for human-robot interaction. The greater affinity for androids and robots has tangible roots in demographics and economics—primarily as a replacement for a shrinking labor pool—but it also grows out of a philosophical tradition that doesn't consider humans exceptional in the ways that Western traditions do. It's no surprise, then, that the leading edge of Japanese AI tends to revolve around humanoid and other robotics, with development of both leading toward what some Japanese experts call "Society 5.0."

The urgency today stems primarily from the demographic cliff the country faces. Low rates of procreation, and an aversion to immigration, have flipped Japan's age pyramid onto its point, leaving companies in a scramble to replace retiring workers. "Labor saving technology of any kind is critical," says Kenji Kushida, a Japan Program Research Scholar at Stanford University's Walter H. Shorenstein Asia-Pacific Research Center. "That's where AI and robotics really come in." However, the latest generation of robots was born into a limbo, caught between two popular conceptions. While robots do far more in 2018 than the traditional factory machines programmed to do one task repeatedly, they remain a far cry from the sci-fi humanoids of Hollywood and Japanese anime.

Still, in this middle ground, developers have made significant gains on the twin tasks of robot perception and grasping, thanks largely to new applications of machine learning. These advances have pushed robotics near a breakthrough in the grasping of arbitrarily shaped objects, such as pine cones, pencils, or wine glasses. "I've been working on the grasping problem for 35 years, and now, with cloud robotics collectively learning from millions of examples, I feel we are getting close to solving it," says Ken Goldberg, head of UC-Berkeley's Center for People and Robots.

Solving the grasping problem would move us one step closer to creating a humanoid robot, but neither grasping nor perception help much in everyday situations if the robot doesn't also possess some basic temporal and causal sense about the environment in which it operates. That sort of cognition includes the kind of reasoning that lets one know that a full mug of coffee needs to remain upright and steady when carried from tray to table, while an empty mug can be flipped and moved swiftly. This level of understanding heuristics—the sort of common sense concepts humans grasp without explicit instruction, or with just one or two tries at a young age—could have broad implications across AI. Yet, they've slipped out of the spotlight, often replaced by brute-force processing on deep neural networks.

The wealth of Japanese research on ways to imbue greater environmental awareness in machines has yet to produce groundbreaking discoveries, but that shortcoming hasn't stopped the country from widespread adoption of automation throughout the country. Stores that normally operated twenty-four hours a day have curbed hours because they can't find enough workers, Kushida says. Many restaurants have replaced cashiers with machines customers use to order and pay. Factories and other skilled trade shops are seeking ways to collect retiring workers' craftsmanship and knowledge and embed it in an AI system that can hold that institutional knowledge and, eventually, pass it along to a new generation of workers.

Fortunately, the integration of more AI-powered robotics and automation didn't require a massive cultural shift in the country, thanks partly to early signals like *Astro Boy*. "When I talk to European colleagues, robots are conceived as some monstrous existential threat that may destroy human society," says Junichi Tsujii, director of the Artificial Intelligence Research Center at Japan's National Institute of Advanced Industrial Science and Technology. "It's a monster type of image there, but in Japan the robot is a protector of human beings or some kind of friend." Because it can do so in a culture that more readily embraces artificial intelligence, AIST focuses most of its efforts on the development of AI systems that will integrate directly in the physical world, particularly in manufacturing, health care, and elderly care. There are limits, Tsujii notes, as Japan clings to its traditions. Yet, even those aren't always sacred. Developers made waves in the country when they taught robots how to perform a traditional dance that was starting to fade away.

In fact, Tsujii and other Japanese AI developers say they believe the general acceptance of advanced technologies derives from something even more-deeply embedded than demographics or pop culture. It emerges from a core philosophical belief that's fairly common in Eastern traditions—that humans aren't as exceptional as they're made out to be in Western thinking. People, animals, plants, and even robots and AI agents reside along a continuum of existence. "We don't have the concept of a creator like God," Tsujii says. "Western civilization is always thinking that human beings are a copy of God and special privileges are given to human beings. Asian cultures don't have that kind of thinking; it's more continuous from animals to human beings." And neither need to be perfect or flawless. In fact, the Japanese have a word for that too: *wabi-sabi*, which includes an appreciation for the little imperfections in everything, from humans to nature to robots.

Yasuo Kuniyoshi goes so far as to stress the importance of lifelike humanoids for the future success of artificial intelligence—for development purposes, but also for integration into society. As director of the Intelligent Systems and Informatics Laboratory at the University of Tokyo, Kuniyoshi designs systems to mimic human anatomy and neurology as closely as possible. "We are trying to build a humanlike thing," he says. "We don't feel it's really an evil thing or a scary thing. . . . That's probably the difference between Western people and Japanese people. Many Western people cannot tolerate a separate existence that's equal to humans."

THE CERN OF AI

Canada shares the Western view, but that mindset—along with the outsized influence of a handful of prominent AI experts—had driven it toward a far less overbearing and more cooperative approach to AI development than in most countries. In a sense, the Canadian ecosystem has become the CERN of AI, building an environment and resources that facilitate mutual discovery much like the famous European Organization for Nuclear Research (CERN) does as a global community research hub for quantum physics. With large corporate actors dominating most of the large data sets in the United States and China, an influential group of developers and investors in Montreal have set out to create an international network of data generators. Canada has a strong start-up ecosystem and commands respect around the world, as do many of its top AI minds, and their openness can help draw more international talent and cross-country connections.

It helps that some of the top AI minds in Montreal, who also happen to be among the most well-known researchers worldwide, share a similar philosophy and push the concept of nonpredatory market competition. Luminaries like Yann LeCun, Geoffrey Hinton, and Yoshua Bengio exert a major influence on the Canadian AI scene. Bengio has an especially considerable impact, says Patrick Poirier, founder and CEO of Erudite AI, which is developing an AI-enhanced peer-to-peer tutoring platform. "Whenever you have a role model who believes and promotes certain traits, you may adopt some of those traits," Poirier says. "I think he has a quite beneficial influence on the market in that regard." On the reverse side, there are rarely big exits for start-ups in Montreal. Maybe if the developers tasted the blood of big profits, they might push harder for it. As it stands, Poirier says, "the community rarely understands or values stock options, and motivation remains more driven by social impact."

Bengio established Element AI, perhaps Canada's best-known AI start-up and one that reflects much of the current mindset in the country. The firm set out to provide an alternative model to the hording of data and talent by large corporations, says Jean-Sebastien Cournoyer, one of the firm's cofounders and a partner at Real Ventures. Before the big Internet titans could swoop in and poach the well-known AI minds and resources in Montreal, Element developed a fellowship program through which top academics could contribute to the company—and get paid quite handsomely to do so—while sustaining their academic research and training the next, critical generation of AI expertise. And with that, it sought to provide a counter to the concentration of AI power in the hands of a dozen or so companies worldwide. "Canadians are collaborative. We're not known to dominate. We're friends with pretty much every country," Cournoyer says. "So, the mindset was, 'Let's build an AI company and build an AI platform, but provide it as a service to all the companies that need it to maintain relevance.'"

By combining the Element platform and the data from its customers, it can help smaller firms create AI to run their businesses. Those clients, in turn, share the knowledge they acquire from data with companies in other industries to strengthen the AI platform as a whole—to build stronger systems for companies that can't build their own. "Our Canadian roots had us thinking about how to build ecosystems that help the world get access to AI," Cournoyer says. In fact, the founders initially considered launching as a not-for-profit, but realized they needed to go the for-profit route to attract top talent and sell the software that helps fund the company. Nine months after Element launched, it pulled in a $100 million round of funding. "For AI to be deployed the way we want as a society, to make us more productive and efficient and grow," he says, "we'll also have to evolve our social fabric and social support system."

But that fabric—the norms implicit in the social contract—can vary widely from one country to the next. The Canadian AI scene envisions a more inclusive future. Japan works toward greater automation. The military helps power innovation in Israel, while Russia continues to grow its heritage of science, if not entrepreneurship. China and the United States connect powerful academics with even more powerful commercial sectors. And yet, the borders between countries can't contain these diverse viewpoints, any more than they can contain the vast flow of data around the world. And so, the grand contest to influence the future of an AI-powered world begins, with the world's powers racing to assert their values, trust, and power.

5

The Race for Global AI Influence

n 2012, representatives from 193 countries gathered in Dubai to hammer out a set of international telecommunications regulations that would govern everything from phone lines to the Internet. Not one of them could've reasonably expected a unilateral agreement on all the issues they would face. The United States would insist on a free Internet with few controls on content, a stance that would surely be opposed by China, Russia, and many other governments around the world. But after some of the preliminary meetings, US Ambassador Terry Kramer felt the delegates might find common ground on at least eight of the ten topics at hand. And then came what he still calls the "oh-shit moment."

In an early meeting to set the foundation for the conference, a Swedish representative argued that the conversation ought to include representatives from NGOs and other civil society organizations. The topic was charged enough and the argument earnest enough that Hamadoun Touré, the secretary general of the UN's International

Telecommunication Union, demanded the Swede's immediate removal. "I believe [Touré] had good intentions in his heart," Kramer says, "but how he led the discussion created a problematic negotiating environment." The signal sent didn't kill negotiations outright, but representatives immediately started to solidify their battle lines and created a "quid pro quo" mindset on issues so fundamental in nature that one couldn't be easily traded off for another.

Still, Kramer saw reason for hope. Despite two issues on which opposed alliances might never find compromise, including content-restricting regulations, virtually all the countries had broad agreement on ideas to fight spam and common cybersecurity threats. Touré might have worked on issues of consensus and built some goodwill, Kramer figures. He didn't, choosing to force votes when more discussion might have produced results. "I think the net effect of his actions was to make certain nations feel 'named and shamed' in a very public setting," Kramer says. "That was a gross miscalculation regarding reactions from nations like the United States and missed an opportunity to drive towards consensus on critical issues that all nations could align around."

It all came to a head on the final evening of the conference, when the ITU leadership allowed a last-minute proposal from Iran to go to a vote without any prior discussion or notice. The Iranian proposal essentially said countries have a sovereign right to regulate the Internet however they'd like. Whether the leadership's decision was an effort to isolate the United States or not, that's exactly what it threatened to do. Kramer hesitantly got up, not knowing whether any countries, including allies in Europe, would support his blanket rejection of any proposal to limit a free and open Internet. The vote was called, and fifty-five nations came out and supported the American position. "The US created a system that allows expression and entrepreneurialism that works," Kramer says. Those supportive countries "were not going to take positions against these key principles."

In the end, the International Telecommunications Regulations, as they were called, were approved by a majority of the countries—the United States in the minority and unwilling to accept or sign the treaty. The Internet has not become free and open around the world, as Kramer and American interests had hoped it would. "This is a long game, and if people aren't up for the long game we're going to have a bad outcome," Kramer says. "You have to hope for success longer term."

Whether in terms of Internet freedom or the regulation of artificial intelligence, the horizon the United States imagined remains well off in a distant future. American and, to some extent, European interests remain solidly in a minority amongst the global community of nations. Almost any way one might try to establish a convening regulatory body—aligned by type of government or by population, for example—and the United States would remain in the minority on these kinds of issues. The one possible exception, to organize by gross domestic product ($1 equals one vote), would represent a vast rift between developed and developing economies. The United States might use the fifty-five countries in its alliance to address issues other countries are advocating, working through other means such as trade negotiations to gain leverage, Kramer suggests. "If you do think there's leverage and a chance for improvement in those other countries, then you try to push that," he says.

Yet, within the community of AI developers itself, some broader consensus seems to be emerging. Concerns about how the world's diverse mix of cultural norms, political needs, and data streams shape AI development has many observers calling for a universally accepted set of standards. IEEE is leading from a technical standpoint with IEEE P7000™ standardization projects, which directly relate to issues raised in *Ethically Aligned Design*, as well as its Global Initiative on Ethics of Autonomous and Intelligent Systems, which focuses on embedding ethical considerations at all stages of AI and robotics development. As John C. Havens coordinates these global efforts, he hopes to push thinking about AI development "Beyond GDP," so we measure success by more than just gross domestic product expansion. "Our goal is to align individual well-being with societal well-being by integrating applied ethical thinking towards new economic metrics beyond growth and productivity," Havens says.* A variety of initiatives around the world are striving toward similar goals, as we discuss in the last chapter of this book, but standards have to go beyond the lab, the work bench, or the corporate boardroom. "We need to stop and reflect before we move into a future in which AI systems affect an individual's agency, identity, or emotion," Havens says. We need a corporate environment in which an engineer blowing

---

*   Interview with the authors via phone, November 21, 2017 the whistle on lazy, ignorant, or nefarious programming is "lauded for bringing innovation into the cycle."

The IEEE almost certainly will adopt a set of standards for its more than 420,000 members in 160 countries. Corporations might even adopt some of Havens's "beyond GDP" thinking about aligning economic growth with human development, as public pressure builds against careless corporate use of so much personal data. But outside the expectations placed on professional engineers and the developing schools of thought about algorithmic ethics, nation-states will continue to vie for supremacy on a geopolitical level. AI isn't just an arms race, although it has very definite military and defense manifestations, as we discussed in the previous chapter. It's a political-cultural race, a battle over cognitive power and its ability to sway mindsets, societies, and economies. The runners in this race include national governments and non-state political actors, but also groups of like-minded individuals, private companies, and other institutions, such as labor unions and educators. Because humans embed their values in AI code, and because we allow those algorithms to make more of our decisions, those systems and the sensibilities of their creators will affect our lives. Will those values come from a group of mostly white and Asian male programmers? Will they come from a central authoritarian government? Or could they develop in a multidisciplinary environment of both private- and public-sector actors that seeks to accommodate the well-being of all humanity?

DIVERGENT PATHS INTO THE AI FUTURE:
THE UNITED STATES, EUROPE, AND CHINA

The race for AI dominance will play out across a few dimensions. The first is *country power* and its different drivers; this includes the amount of funding provided to scientists and entrepreneurs, the collection of scarce AI talent, the caliber of research and the fluidity of the entrepreneurial ecosystem, as well as the ability of a country to unify and outwardly project its civil society's value system and trust. The second dimension builds on this, but blends in *individual power*, as well. This is driven by the ability of citizens to exercise and change their personas and choices freely, the ability to take recourse in the face of mistakes made by AI systems, the existence of off switches or opt-out paths, and a way for citizens to help shape AI governance, all without stifling its growth—a difficult balance, to be sure. Finally, there are drivers of *institutional power* that will shape AI development: Are data sets sufficiently large, statistically valid, and accurate, and do they comply with local norms of interest representation? For instance, do programmers, companies, and governments have the resolve to codify systems in a way that balances individual and community goals? Do they safeguard privacy, agency, and the power to protect one's true persona? Is there sufficient technocratic expertise and capacity to govern AI transparently on behalf of their citizens, facilitating growth and safeguarding abuse?

This mix of country, individual, and institutional power is hardly new, of course. It resides at the center of most political and economic interactions today, and has since the inception of social contracts and nation-states. Yet, as Henry Kissinger notes in his essay "How the Enlightenment Ends," AI fundamentally changes these inherently human interactions. Previously, our interactions forced us to ponder our interpersonal and institutional relationships, reflecting on our values versus theirs, training our critical thinking capabilities, and honing our creative skills to improve these partnerships. Artificial intelligence relieves us of some of those burdens, adding great convenience but also, if we're not careful, a numbing of deep consideration and decision-making abilities.

Individuals and institutions will have to evolve new ways to balance efficiency and convenience with the need for an educated and civically fit citizenry. Many people and organizations will stumble away from this fine line, readily accepting the ease and convenience of better technological tools and avoiding the arduous and lonely process of deep reflection. (We already see this happening on social media, in ways that change the actual neurological wiring of our brains.) Pressure will grow for institutions to lower hurdles and create more fluid channels for data, opening the pipeline for the types of relationships, investments, and pronouncements that make for sensational headlines and play to our basest instincts. Yet, despite that, a truer power might begin to emerge from institutions that balance the convenient with the conscious, focusing as much on human growth as economic growth.

---

* Henry A. Kissenger, "How the Enlightenment Ends," *The Atlantic*, June 2018.

Countries that help facilitate this balance will attract the companies and people who appreciate the value of thoughtful design, patient deliberation, and a search for the common good. This does not necessarily point to an advantage for democratic governments or free-market economies. One can already see the nascent stages of such deliberate approaches in a diverse set of countries, from Denmark and Sweden to Singapore and the UAE. Regardless of whether an outsider finds their philosophies agreeable or objectionable, these countries support a coherent philosophy of cognitive growth. They're instituting a unified approach with a high degree of technocratic support, sophisticated planning, and a scientific and technological competency that, as Parag Khanna points out in his book *Technocracy in America*, facilitates the growth of economic and political power in today's world.*

People, institutions, and countries will disagree with one another's approaches to these dimensions, especially when it comes to regulation and data-protection regimes. Nations have battled over these fronts throughout the last wave of globalization, whether about free speech on the Internet, free passage for commercial airline traffic, restrictions on ownership of telecommunications providers, visa regimes for immigration, tariffs on trade, or taxes on multinational corporations. We can already see some even more formative and powerful differences in political and strategic directions taking shape around artificial intelligence, and each of them will have a major impact on the cognitive race ahead.

The three major AI powers have followed their history, hearts, and minds into the future. The United States approach features the "Brawny Barons," relying heavily on its free-market capitalism and the strong private sector and entrepreneurial community it produced. While critical funding for basic and advanced research comes from government sources, the bulk of innovation and control rests in the hands of the country's start-ups and giant digital companies. In China, of course, the government and the Party take a more direct hand in the development of advanced technologies—the "Party Protectorate" in place since Communism took hold. The country has its powerful

---

* Parag Khanna, *Technocracy in America: Rise of the Info-State* (CreateSpace Independent Publishing: January 2017).

Digital Barons, too, but the lines between corporate and government influence have blurred, and in some cases have been eliminated, after many years of government letting the Barons run their own show. Western Europe is trailing both the United States and China, but recently chose a middle path: a "Community Commons" in which its lack of Digital Barons and a measured level of governmental involvement has led to a collective EU-wide effort to balance innovation and individual protection.

THE EUROPEAN UNION: COMMUNITY COMMONS

As we discussed in the previous chapter, Europe has no big Digital Barons of its own. Having sought multiple times to limit Google and Microsoft through antitrust proceedings and possessing all too much historical experience with totalitarianism, the European Union feared that US or Chinese companies might mine citizens' data and keep the information in jurisdictions traditionally less concerned with individual protections (if for different reasons). Of course, privacy issues concern more than just Europeans. Chinese citizens pushed back on systems that encouraged neighbors to report on one another. And Americans have paid more attention since learning that the National Security Agency spied on "persons of interest" within its own borders after the 9/11 terrorist attacks. In fact, private-sector companies have heightened those concerns as they've reported a growing number of data breeches or have actively shared personal information in ways that stretch the bounds of their users' trust and privacy. Facebook and its data-sharing agreements with Cambridge Analytica and other partners, including Chinese companies, has landed CEO Mark Zuckerberg in the glaring spotlight of US congressional hearings, setting up a confrontation with which neither side is comfortable.

To its credit, the EU has moved a step beyond other countries and regions in this regard, tying individual protections to a more stringent set of requirements for data and artificial intelligence. The formative General Data Protection Regulation puts more agency over personal data in the hands of the individual. If someone wants their data deleted, a company has to do so or face significant fines. The rules also require that companies have the ability to explain why their systems made a particular decision about a particular person. They need "explainable AI," a difficult proposition because, as we described, many of the complex neural networks that enable machines to learn can't tell human operators how or why they reached their conclusions. Making AI explainable seems generally desirable, but it brings certain important drawbacks. From an economic standpoint, it could destroy competitive advantages for commercial designers of these neural networks and therefore reduce investment in useful applications of them. It could also lead to the development and application of only very narrow types of explainable AI systems, while more powerful or beneficial applications work in less restrictive markets. The early regulation could sharply limit experimentation.

Damian Borth at DFKI notes that many researchers are torn by the regulations, understanding the clear need for individual protections while lamenting the chill it could put on AI development and deployment across the EU. "We don't know everything about how an airplane works before we fly in it," he says. What's most critical is that we have assurances that the aircraft, pilots, and air traffic control will work effectively and safely. But that, too, requires a certain threshold of regulatory oversight, to ensure that the dire scenarios don't emerge and potentially trigger an even sharper backlash.

Governments and regulators need to address the difficulties, including personal privacy issues and the threat AI poses for labor, says Catelijne Muller, a member of the European Economic and Social Committee and rapporteur on artificial intelligence. EU officials have brought together different interest groups in society to discuss the impact on jobs, with labor union officials collaborating alongside corporate executives to gain a better understanding of possible futures. "If we want to benefit, truly benefit from all great potential of this technology, we should address the challenges," Muller says. "If we don't address the challenges that are obviously there, one government in the future is going to say they're going to prohibit this. It's gone too far. So, I don't think of this as stifling innovation, but promoting it in a sensible way."

An EU report released in April 2018 set out a "European approach to artificial intelligence and robotics."* The plans included a sharp

---

* Robotics and artificial intelligence team, *Digital Single Market Policy: Artificial Intelligence*, European Commission, Updated May 31, 2018.

expansion of annual investments in AI by 70 percent under a framework program called Horizon 2020. The commission said it would increase its own funding to $1.8 billion (EUR 1.5 billion) between 2018 and 2020. If member states and the European private sector make similar efforts, the report says, the Commission believes this number ought to grow to $24 billion (EUR 20 billion) by the end of 2020. The EU countries and private-sector entities should then aim to invest that same $24 billion amount *each year* in the decade after that, the report says.

The report also set out plans to create 500 digital R&D centers and connect existing research centers around Europe, all of which would support AI development and prepare for the socioeconomic changes it will bring. This will be flanked by an ethical and legal framework to provide more clarity on EU member countries' expectations for AI use. However, nothing in the directive hinted at a seedbed from which a new Digital Baron would grow to accelerate AI development in the private sector. Instead, the strategy lays out what might become an alternative to the Digital Baron model that helps drive the AI economy in the United States and China.

With what looks more like an "AI On-Demand" strategy, the proposed EU approach would treat AI like a basic infrastructure investment for the digital economy, in a vein similar to power lines or fiber-optic cables. It goes further, however, adding two additional steps that look more like green energy initiatives. First, the plan includes efforts to democratize access to new AI models and a large depository of data. That would give individuals and organizations of all sizes access to the two primary ingredients for development of new AI applications, not unlike the Element AI initiative in Canada (mentioned in Chapter 4). Second, the EU plan contemplates the formation of a couple new funding mechanisms: a $605 million (EUR 500 million) "European Fund for Strategic Investments" to support development and uptake of AI; and a $2.5 billion (EUR 2.1 billion) pan-European venture capital fund-of-funds program. That money could prove especially vital in the global AI race. According to McKinsey data referenced in the April 2018 EU report on AI strategy, Europe lagged other regions' private investments in the field, with around $3 billion to $4 billion in 2016. During the same year in North America, private investment in AI totaled

$15 billion to $23 billion. Asia's investments ranged from $8 billion to $12 billion.*

These proposals would mark a clear step forward for the EU and send a strong signal of conscientious development and harnessing of AI for social and economic growth. But Europe is clearly still a step behind the kind of spending it sees when it gazes east and west. It also needs to prove out that its digital single market, safety, and cybersecurity frameworks will succeed in creating one contiguous market of 500 million people for AI innovation. That has been a historical challenge in a region that is tremendously diverse. While that variety could become a potential strength from a data perspective, in many ways the region remains economically, politically, and culturally fragmented. That makes it hard for entrepreneurial businesses to expand. Size still matters in markets and economics.

Ironically, though, the existence of the EU's digitized welfare state might also open the door for more AI development in Europe. For instance, Maja Hojer Bruun, a professor and techno-anthropologist at Aalborg University ran an experiment to gauge Danish residents' reaction to drones. Bruun and her colleagues flew drones over people's homes and then interviewed them while the drones were overhead. Almost all assumed the government would have proper regulations in place, and operators would follow those rules. Not all European countries and people would share such a relaxed attitude, which appears to stem in part from Denmark's legacy of broadly digital government services, Bruun says. But this is an example of the fact that many Europeans trust government stewardship of disruptive technologies.

Denmark is ahead of the pack. The Danish government has a positive track record when it comes to building digital trust with its citizens. Years ago, it put in place reasonably far-reaching digital platforms, such as Citizen.dk and citizens' electronic mailboxes ("Ebox"), which let people submit taxes and consult other government services online. Denmark also requires schools to teach computer science and discuss digital culture, which builds digital competencies in its children. So, while Bruun still has concerns about the interpretation and reduction of individuals when they're being digitized, she

---

* James Manyika, "10 imperatives for Europe in the age of AI and automation," McKinsey & Company (October 2017).

has little doubt that most Danes will trust AI innovations that are regulated by the government. "There's a high degree of trust in the authorities," she says, "but it is very important to maintain this trust and not to jeopardize the relationships of trust by selling people's data for commercial purposes. People are willing to improve, for instance, public health and infrastructure with their data, but they always want to see a meaningful purpose for themselves as users and citizens or for society, and not just participate for the benefit of companies."

And indeed, some European countries, especially Sweden, post high digital trust ratings, according to a report by Bhaskar Chakravorti, a senior associate dean at the Fletcher School at Tufts University.* Chakravorti and his colleagues researched digital attitudes, behaviors, environments, and experiences across 42 countries, scoring the nations with a trust rating in each of those four categories. Looking at attitudes, or "how users feel about the digital environment," France and Norway (each at 2.41) and the United States (2.45) posted lower digital trust ratings than Pakistan (2.66). Germany was only marginally higher (2.73). Compare those with China, at 3.04. Having broad access to digital services doesn't translate into trust in those technologies. And while there's no empirical evidence that low trust translates into low national power in the global digital economy, it stands to reason that it doesn't help. Clearly, China's ability to generate prosperity based on technology development, despite all party interference, creates a more supportive populous for digital policy.

Of course, for all its ability to reflect critically on these new digital trends and build protective moats around its citizens, Europe lacks critical mass when it comes to digital global players. It does not have a viable alternative model to techno-economic development that the world could latch onto. Berlin, Hamburg, London, and perhaps soon Paris have vibrant tech scenes, but they have not yet scaled out to regional, much less global mass markets, aside from a bit of digital penetration in manufacturing. Will European countries and the EU seek to build out this AI economy, or just circle the wagons, and can they influence the United States, China, and the global market if they

---

* Bhaskar Chakravorti, et al., "The 4 Dimensions of Digital Trust, Charted Across 42 Countries," *Harvard Business Review* (Feb. 19, 2018).

only do the latter? "I worry about Europe's economic development," says Ambassador Kramer, who lived in Europe for many years before receiving his political appointment in the United States. "If you can't bring scalable digital assets to the global competitive environment, how will you grow? How will you be able to negotiate and have influence in the manner you'd like to?"

RUSSIA: RATTLING THE SABER, BUT WHAT'S REALLY THERE?

While Western Europe grapples with the balance between AI innovation and individual agency, their towering neighbors to the east appear to have few such concerns. The rhetoric out of Russia suggests a singular focus on the role of AI in asserting geopolitical power. Yet, beyond President Putin's pronouncement that the country that leads in AI "will become ruler of the world," the underlying ecosystem for development appears to lack some of the key assets necessary for broader AI leadership, at least on the commercial side.* Russia still flexes plenty of muscle in the realm of security. "Don't underestimate the deep Russian science and technology expertise," says Horst Teltschik, former national security adviser to German Chancellor Helmut Kohl and a long-standing expert on Russian affairs. Teltschik points back to 2004–2005, when he and I (Olaf) worked together at Boeing International's German operation, where he was president. At the time, our Russian colleagues could boast the creative power of 2,000 engineers in Moscow, a tremendously important, cost-effective asset for the American company. "It is true that around 100,000 young Russians leave their country every year to seek brighter career horizons elsewhere," Teltschik says now, "but those that want jobs in the space or defense industries have very good prospects. After all, America is using Russian rockets to carry payload to the joint space station."

Indeed, state-sponsored technology development is alive and well, in no small part because Putin's primary interest lies in the maintenance of autocratic power, explains Evelyn Farkas, former US deputy assistant secretary of defense for Russia and Eastern Europe. Putin

---

* James Vincent, "Putin says the nation that leads in AI 'will be the ruler of the world,'" *The Verge*, Sept. 4, 2017.

rode to heightened power on the wave of oil and gas production, but those commodities were volatile and couldn't reliably secure citizens' allegiance. So, officials started to close the "tricky but open" environment, forcing the departure of NGOs and companies such as LinkedIn, Farkas explains, and the government began to demand back doors into servers. Yet, compared with China, the environment remained fairly open to Western Internet companies—so the idea of iPavlov receiving government funding to create an open-source database for language processing, as noted earlier in this chapter, made sense.

Yet, despite the rise of "Novo Rossiya" and its renewed emphasis on empire and a reinvestment in the pride of the past, Russia's science and technology capability has deteriorated, Farkas says. The country no longer enjoys the breadth and depth of research as it did during Soviet days. While this could shift with Putin's emphasis on artificial intelligence, Russia has displayed little interest in research and development, leading to a brain drain as young professionals head to more supportive environments, such as Israel. One of the reasons for those departures is the difficulty scientists have transitioning between academia and science, explains Mike Kuznetsov, a consultant for Aspera, an IBM-owned company, and a Russia strategist at Cambrian.ai.

One of the few ways to get recognition among the glitterati of AI development in China and the West is to publish findings openly and participate in open-source development. But to the extent that researchers in Russia can participate in commercial projects, they typically have to partner with the large, government-owned entities such as Sberbank or Gazprom. "Scarcer entrepreneurial opportunities leave many scientists choosing between research projects that benefit one of the government corporations, staying in academia and basic research, or just making less money from grant funding," Kuznestov says. That limits the possibilities for bringing cutting edge new insights and technologies to disruptive new ventures.

CHINA: THE DIGITAL DRAGON STARTS TO ROAM

Like most world leaders, Putin seeks to extend his country's geopolitical influence, in some cases by disrupting the political and societal cohesion of other countries (Exhibit A: the 2016 US presidential election). China seeks to do all that and more under the leadership of Xi Jinping. To Xi, the Party, and most Chinese residents, artificial intelligence represents a vital engine that will drive the country back to its historical status as the greatest society on the planet. The spreading influence will ride on the "Belt and Road Initiative" (BRI), an ambitious plan to extend China's reach through billions of dollars of infrastructure investments in developing countries. With this bold emergence from its past isolationism, China aims to reassert itself as a dominant force on global power and culture. AI plays a vital role in that, riding on top of the infrastructure China builds out on land and in seaports across much of Asia. The BRI aims to recast the old Silk Road trading route to Europe and a chain of seaports that secure China's access to trading hubs and energy exploration throughout Asia, Africa, the Middle East, and Europe.

The race to lead in artificial intelligence both supports and illustrates those ambitions. The idea of China reclaiming its rightful place in the world after what many Chinese people regard as centuries of exploitation by foreign powers has become a matter of great national pride. The Communist Party has fomented a powerful narrative about the rejuvenation of the great Chinese people, and those people—especially the educated and elite, but increasingly the general population—are seizing on artificial intelligence as a primary tool to see that through.

As such, AI-powered technologies play multiple roles, which explains some of the country's mindset, including its views on privacy that differ from the United States and Europe. "That's not a Chinese leadership thing, but a Chinese cultural thing," says Amy Celico, a principal at Albright Stonebridge Group, a global strategic advisory firm that helps clients navigate the complexities of international markets. That doesn't mean that the government or the citizens don't care about privacy—quite the opposite. China passed data privacy laws in 2017 that ruffled feathers in the United States because they required storage of data within the country, but the regulations were intended in part to help limit the sorts of commercial intrusions that are bothersome but common in America, Celico explains. From a US perspective, it looks like the Party controls data so they can track it. However, through a Chinese lens, the party is preserving the safety of that data. "The government cracking down on privacy is not to stop dissidents, but to get more control over society," she says.

That stability is paramount. A 2013 study led by Harvard University social sciences Professor Gary King found that Chinese censors are worried about social order, not criticism of the party or government. He and his colleagues scraped posts on almost 1,400 social media services before censors could remove objectionable material. "Contrary to previous understandings, posts with negative, even vitriolic, criticism of the state, its leaders, and its policies are not more likely to be censored," they wrote in their paper. "Instead, we show that the censorship program is aimed at curtailing collective action by silencing comments that represent, reinforce, or spur social mobilization, regardless of content."* China's burgeoning advantages in AI help preserve that stability, feeding a national pride and preserving support for the Party.

The country still has mundane and structural challenges that could slow its progression, as Celico notes. AI might help address some of those issues, but in itself it's not sufficient to solve them. For example, the government continues to struggle to provide basic health care across such a massive population, so it must balance its spending on AI and advanced technologies with initiatives to improve its comparatively low levels of basic medical care. The country also continues to cope with the wave of rural migrants moving to urban centers in search of economic opportunity. Many public services, including health care and childhood education, are tied to the country's *Hukou* system of household registration, which identifies residents' hometowns and other personal information. The government bases many of its benefits on rural or urban residency status and, for a rural migrant, establishing a Hukou status in an urban area can be difficult, especially in the largest cities. Without an urban Hukou, parents who move to the city have to pay to send their kids to schools that are free for urban residents. While the government has made some changes to the program, which essentially creates an informal caste system, it still could leave millions of Chinese with less access to key government services.

---

*   Gary King, Jennifer Pan, and Margaret E. Roberts, "How Censorship in China Allows Government Criticism but Silences Collective Expression," *American Political Science Review* 107, no. 2 (2013).

China has the necessary resources and tools to address these issues, and to accelerate its push into artificial intelligence, robotics, semiconductors and life sciences at the same time. Going back thirty years, only the United States had the capital, market, people, and technology innovation to lead the world. "China now has all of those as well," says Ya-Qin Zhang, president at Baidu and former head of Microsoft Research Asia. "The talent is here. The technology is still behind the US, but the gap is narrowing. The market and the capital here are as good as the US and could be an advantage given the population scale." But the country also has a significant advantage, what Zhang calls "the China speed." Chinese people are very open, even in traditional industries, to new ideas. Stores already are eager to use AI, even if they don't know what it is. Almost every type of consumer transaction has switched to electronic payment methods, rather than cash or credit cards. And surveys suggest some 90 percent of the population supports driverless cars, compared with 52 percent in the United States, he says. No doubt some of this sentiment derives from the extraordinary traffic congestion and population density in China's megacities, not to mention the pollution in most of those urban centers, but the Chinese populace also expresses a greater willingness to embrace a range of new technologies.

The speed comes in part from the government's ability to mandate and often implement sweeping initiatives and from its eagerness to enlist the country's large corporations in those plans, such as the use of renewable energy to lessen pollution or the social credit system to increase trust and commerce. That happens at levels the US government would never broach with Microsoft, Amazon, or other private-sector titans. Yet, Zhang says, the wellspring of that speed advantage still flows from the Chinese people themselves. They've witnessed the emergence of China as a high-tech power in recent decades, and they revel in the success of the country's leading-edge global brands. But perhaps even more importantly, Zhang says, they see a consistent direction from the Chinese government and believe they can tap into some of that tech-fueled prosperity, too.

Jack Ma exemplifies this as much as any Chinese citizen can. Ma failed his primary and middle school exams multiple times, did the same with his college entrance exams, and then struggled to find a job after finally graduating. He was the only one of twenty-four applicants to be rejected for a job as a KFC manager. As he tells it in multiple interviews, he applied to Harvard ten times and never got in. He struggled to get venture capital funding for his new company, which he founded in 1999.* By 2018 that sad little start-up, called Alibaba, had grown into one of the largest digital companies in the world. As of April that year, Ma was worth an estimated $38.5 billion.†

THE UNITED STATES: LEADERSHIP IN THE BALANCE

In 1987, the eminent economist Lester Thurow delivered an address in Sendai, Japan, and one of his remarks stuck with me (Mark) over the many years since. Thurow noted that, after World War II, the United States led the world in every industry except one: bicycles. Italy took that solitary honor at the time. By the time we wrote this book, that undisputable pole position dwindled and gave way to a more balanced, multipolar global economy, including in some advanced technology fields. The rising geopolitical and technological muscle of China has not yet diminished US leadership in artificial intelligence. American universities, companies, and government-backed initiatives continue to push the frontier of innovation, and the emergence of competing powers still springs, at least in part, from the cutting-edge work done in Silicon Valley, Boston, Seattle, and other US high-tech centers. According to the Chinese Ministry of Education, more than 608,000 students left the country to study overseas on 2017, most of them going to universities in the United States and Europe. The number of students returning to China, often dubbed the "sea turtles," increased more than 11 percent in 2017—and of those, almost half had earned a master's degree or higher from Western universities that are still considered the cutting edge of breakthrough thinking, both in terms of science and technology and their socially critical examination.‡

The Stanford Center for Advanced Study in the Behavioral Sciences is just one such leading center. After several years leading DARPA

---

*   Ali Montag, "Billionaire Alibaba founder Jack Ma was rejected from every job he applied to after college, even KFC," CNBC, Aug. 10, 2017.

†   Fortune 500 profiles, "No. 21: Jack Ma," *Fortune*, Updated July 31, 2018.

‡   Ministry of Education of the People's Republic of China, *2017 sees increase in number of Chinese students studying abroad and returning after overseas studies*, April 4, 2018.

during the Obama administration, Arati Prabhakar moved back to Silicon Valley, where she had been a venture capitalist, to take up a fellowship there. Through the fellowship she's pushing the outer limits of how we model and understand the extreme complexity in today's world. Much of her advanced research grows out of a fertile blend of experience that's not uncommon among America's innovation leaders, and it makes her ideally suited to help drive the story of US science and technology leadership in AI. For example, she's contemplating the concept of "adaptive regulations" that "allow you to experiment and learn without going too far," she says at a coffee shop just off campus. She notes that policies and regulations should achieve a degree of consensus and then provide stability so individuals and companies can count on a set of ground rules for a certain period of time. "We won't ever make the pace of regulation as fast as the pace of technology—we would be whipsawed if we did," she says, "but we can keep it closer."

Advanced technology, including AI, could have the greatest impact on humanity as it tackles societal problems and learns about human behavior, Prabhakar says, but "we're really, really early on that." While cognitive machines can process staggeringly large arrays, none can bring any depth of understanding to the table. So, researchers try to build economic and behavioral models that can leverage the massive but narrow computational power of AI in a way that lends to better human understanding. But, of course, models are inherently inadequate, so we add and innovate on top of them to make better models. "If you want to take those next leaps to have richer, more representative models—knowing that you're not going to emulate everything—how do you do that?" she asks.

At one point, DARPA worked on a program to develop a model that might predict food crises in places like Africa or the Middle East, tracking weather, soil conditions, and several other environmental and human factors. Yet, one could never fully model how a government regime might react, and thus jeopardize or facilitate agricultural production. So, a deeper model will have to factor in those variables as well. "The overarching narrative with IT today is the ability to tackle scale and complexity we never thought was possible before," Prabhakar says.

In this sort of complex-systems thinking, beyond just the deep technical and research expertise dedicated to AI development, American researchers remain better at defining the "future evolution of AI," says Tom Kalil, chief innovation officer at Schmidt Futures and the former deputy director for technology and innovation at the White House Office of Science and Technology Policy (OSTP). Yet, a clash of models has ensued, and it's not entirely clear what will emerge as they interact. "China's political economy has a greater focus on maximizing national power; America's political economy is better at the efficient allocation of capital," Kalil says. "If China is willing to spend whatever it takes to establish a leadership position in technologies such as AI and quantum computing, it may be inefficient but could still be effective. I don't think America's political and business leadership have a strategy for dealing with this."

China still has major institutional challenges and issues with academic performance and integrity, but the return of the "sea turtles" will help mitigate some of those issues. And upon their return, they bring new technical, managerial, and cultural insights that will enhance China's AI ecosystem. To the extent that AI breakthroughs will benefit humanity in some ways regardless of their country of origin, this intersectional relationship between the United States and China and other countries has the potential to improve lives around the world. What begins to trouble some American experts, however, is the degree to which China fuses civil society with defense objectives. A clear divergence in values is emerging, values that govern whether and how regular citizens should contribute to national security. The United States, and Western militaries in general, draw a clear distinction between war and peace, and the separation between civic and military domains whereas China's People's Liberation Army tends to see them on a continuum.* Political or economic competition is viewed as part of an ongoing struggle in which every citizen plays a part, even if the country remains far from any outright military clash.

This provokes a great deal of concern about China's interference in Western democracy and society, or the poaching of western intellectual property in otherwise unassuming business or personal relationships. That's why it is important to distinguish between the traditional concept of an "AI arms race" and the intelligence and counterintelligence operations that countries conduct toward one another on an ongoing

---

* Elsa Kania, "China's quest for political control and military supremacy in the cyber domain," *The Strategist*, Australian Strategic Policy Institute, March 16, 2018.

basis these days, says James Andrew Lewis, senior vice president at the Center for Strategic and International Studies. In the race for AI leadership, Lewis says, "military terminology doesn't make sense. It's not a war. It's not an arms race." The push to drive greater digital innovation in the military spheres is hardly a novel concept. However, he says, the United States very definitely faces a new, politically divergent competitor for economic and cultural influence, a counterforce it hasn't seen since the Cold War and one that sees the rivalry as a competition across multiple domains, all to be won or lost.

The Office of Science and Technology Policy under President Donald Trump adopted a more libertarian approach to AI-related regulation, both domestically and internationally. Within the United States, the president's Office of Science and Technology Policy will seek to reduce barriers to high-tech start-ups, aiming to keep the country at the forefront of entrepreneurialism and innovation, according to public remarks by Michael Kratsios, the deputy assistant to the president for technology policy. In a May 2018 address announcing the creation of the National Science and Technology Council's Select Committee on Artificial Intelligence, Kratsios said that, in many cases, "the most significant action our government can take is to get out of the way."* The administration would not try to solve problems that don't exist, he said. Rather, it would seek to provide the private sector with more access to resources such as government labs and data. And while Kratsios also partnered with other G7 representatives to declare "the importance of investing in AI R&D and our mutual goal to increase public trust as we adopt AI technologies," he also stressed that the White House "will not hamstring American potential on the international stage." Domestically, command-control policies can't keep up with private innovation, and the administration will not bind the country "with international commitments rooted in fear of worst-case scenarios," he said. "We didn't roll out the red tape before Edison turned on the first lightbulb."

As we explore in Chapter 8, reliance on existing international institutions might also result in a lack of expertise and a lack of inclusiveness that could drive Chinese authorities to create their own model of

---

* *Summary of the 2018 White House Summit on Artificial Intelligence for American Industry,* The White House Office of Science and Technology Policy, May 10, 2018.

international governance. As it asserts itself, China no longer wants to play under Western rule sets and regimes, and the rulebook for both hot and cold wars has changed. During the Cold War, we dealt with an adversary proficient at psychological warfare, and we certainly had our own influencing techniques, such as Radio Free America and the creation of universities in West Germany and the Middle East. Since then, we have conceived of "psych ops" as the business of Fifth Avenue rather than Pennsylvania Avenue. "An Egyptian colleague told me, 'You Americans, you're hopeless at propaganda,'" Lewis says. "'You think it's like selling soda pop.'" That might be giving the United States too little credit, but it's clear China and Russia have recently posted a strong track record of multiplying psychological impact through AI and social networks, which bind to our emotional receptors much more effectively than leaflets or in-person seductions. Today's battleground is the deep tissue of society in what is now called a "hybrid conflict" through civil-military fusion (CMF).

Today, Chinese and Russian military doctrines consider cognitive and economic actions, seeking to gather more public and private data or sow confusion in rival societies. "The big changes will be economic," Lewis says. Efforts to integrate AI and autonomy into weapons systems has gone on for years, so more intriguing now is the fact that "your decision making will change as a consumer or business because of the ability to access AI." To wit, the United States Cyber Command has shifted its perspective on cyber activity from the idea of individual hacks or attacks to one of sustained, sophisticated campaigns to undermine anything from American military power to social cohesion.[*]

Of course, that doesn't mean AI in traditional defense applications doesn't matter. While China has made great advances in its cultural and economic influence, the United States has retained, for now, an edge on power and smart military technologies. This emerges in part from the Third Offset Strategy initially laid out in 2014 by then secretary of defense Chuck Hagel. The strategy seeks to lead the integration of autonomous and other AI-powered technologies into "warfighting potential" and restore the military's "eroding conventional overmatch versus any potential adversary, thereby strengthening conventional

---

* Richard J. Harknett, "United States Cyber Command's New Vision: What It Entails and Why It Matters," *Lawfare*, The Lawfare Institute, March 23, 2018.

deterrence," the Hon. Robert O. Work, himself a former deputy secretary of defense under both the Obama and Trump administrations, writes in a report on US Department of Defense spending on advanced technology.*

The report, produced by Govini, a government analytics and big data firm where Secretary Work is a board member, found that unclassified defense spending on AI, big data, and cloud technologies reached $7.4 billion in fiscal 2017, up by 32.4 percent from fiscal 2012. While artificial intelligence accounted for just a third of that total, it was the largest contributor to the increase over that five-year span. Major flows of funding went to virtual reality, virtual agents, and computer vision, explains Matt Hummer, Govini's director of analytics and advisory services and coauthor of the report. The most growth centered on intelligence, surveillance, and reconnaissance, activities that generate massive inflows of audio, video, and other data that can be sorted through and parsed by AI-powered technologies. In one DARPA program Hummer describes, natural language processing programs work in conjunction with virtual agents to provide advanced field translation services for the niche dialects soldiers might encounter while deployed overseas. However, now, the US military could record even harmless chats with unassuming civilians, evaluate them, and then use them to direct military strikes, inadvertently turning civilians into informants and putting bull's-eyes on their backs. In other applications, the immense amount of video footage gathered by reconnaissance drones can now be evaluated in a much more holistic fashion. Whereas human analysts previously struggled to identify all objects of interest in a given frame, machine learning can now analyze contextual images much more quickly and effectively. This is a powerful advantage for a nation that has more military data from operations in recent history than any other on earth. Both the United States and Chinese private sector have built the huge data sets necessary to train most AI models. But when talking about military applications, it's not just any data set that matters. "It's collecting data in operating contexts," Hummer says, "and the US has a huge advantage in those spaces."

So, defense spending on both sides won't decrease in these advanced fields any time soon, especially considering that the results of

---

* *Artificial Intelligence, Big Data and Cloud Taxonomy*, Govini, 2017.

innovations at DARPA and similar agencies can and often do trickle out into commercial use over time as well. Continuing to push from a leading position on the frontier will help retain both a military and economic edge. And for this, as Hummer points out, big data remains essential.

THE COMING CLASH OF MODELS

This will not be a two- or three-horse race. While the United States and China clearly lead the AI competition, and a well-established second tier with the United Kingdom, Russia, and Israel are not so far behind, dozens of other countries will play a key role in a future of pervasive AI deployment. The United Arab Emirates have created a government ministry solely dedicated to artificial intelligence, and the Emirates have opened their arms to companies seeking to test ideas like personal drones for transportation and other futuristic advanced technologies. In October 2017 the vice president and prime minister, Sheikh Mohammed Bin Rashid Al Maktoum, appointed twenty-seven-year-old Omar Bin Sultan Al Olama as the first state minister for AI, charging him with the task of making the UAE "the most prepared country for AI" through the "pursuit of future skills, future sciences, and future technologies." What path the UAE will take exactly—whether regulation-forward like Europe, experimentation-forward like the United States, or decree-forward like China—will reveal itself in the years to come. What is clear, however, is that AI will become a more holistic economic-development strategy that aims to establish the Emirates as a hub for futuristic experimentation and investment, including in ideas such as the Hyperloop, a high-speed transportation concept popularized by Elon Musk; autonomous passenger drones being tested by Chinese micromultinational eTang; or new projects in desalination and solar energy. Having learned the lessons of its meteoric rise onto the stage of the world economy thanks to oil and high finance, commensurate with their respective boom and bust cycles, the Emirates are now diversifying into the future. This will be further fueled by competition with neighbors like Qatar and Iran, both technologically advanced Middle Eastern powers with rivaling economic positions, political interests, and allies in the region.

Small, culturally more homogenous, reasonably cash rich, and with a solid business infrastructure, the UAE already meets some of the accelerating conditions for AI investment, not unlike Singapore or Denmark. Its centralized government and emphasis on safety, security, and stability over individual privacy and freedom make it, like China, an open field for experimentation. However, it lacks a rich tradition and ecosystem of digital entrepreneurship inherent in the United States, and, as a small nation, the large data pool required to train AI systems. But those mixes of advantages and disadvantages are limiting factors, not outright barriers. Anyone with the right data set, the right expertise, and the right amount of computing power can leap ahead in this race, and their participation will start to reshape the geopolitical doctrines we understand today.

As Parag Khanna, author of *How to Run the World* and *Connectography* suggests, we're in a new version of the medieval world, where myriad actors—including governments, cities, corporations, NGOs, and individuals—negotiate for power and influence. It remains to be seen whether we turn that into a new renaissance or another global conflict. That was the picture before the new AI spring, too, but data now flows silently across borders, and what's collected in one country might be processed and audited by a small cadre of highly qualified engineers in another. Those who can attract and help grow that cadre could end up creating a vibrant new renaissance.

Even if their intentions are noble, rarely does that group of entrepreneurs or researchers include experts who can draw conclusions about the cultural, ethical, or legal implications in different countries. So, national governments continue to experiment with different types of AI regulation and policy without much concern for the often invisible digital fallout. As we have explained, some are predisposed to a laissez-faire approach, while others will take a more proactive role in regulating the use of personal data and the transparency and explicability of AI processes. Others simply rule by ad hoc degree. This divergence will become more pronounced in the decade to come as the Digital Barons seek to expand their multinational reach in their insatiable drive for data and profit. This will heighten old concerns and generate new ones. Philosophies of regulation, influence, and social and economic participation will conflict—as they should.

Those clashes and their outcomes will coalesce around issues of values, trust, and power. A sustained assault on US society and institutions might prompt the government to militarize its citizens, raising new questions about America's conception of its moral authority and exceptionalism. Artificial intelligence has moved beyond cyberwarfare and now interferes with the individual lives that make up the fabric of societies. That could threaten America's deeply valued separation of military and civilian spheres. Notions of trust also will come into conflict, with the United States and Europe likely to lead the development of AI regulatory models that preserve individual control of data and how it's used. While China takes a different path on individual privacy, Western companies and governments might need to develop new business models that account for much greater individual control and agency—and thus enhance trust.

That trust could play out, then, in the race for geopolitical influence and soft power, with Western countries offering trust-based models but less investment, and China offering more money and infrastructure but less individual control over data and civil rights protection. As Kai Fu Lee, the CEO of Sinovation Ventures writes in a *New York Times* op-ed, that could make developing countries an "economic dependent, taking in welfare subsidies in exchange for letting the 'parent nation's' A.I. companies continue to profit from the dependent country's users. Such economic arrangements would reshape today's geopolitical alliances."*

Meanwhile, we should not assume that US science and technology leadership in cognitive computing and related applications cannot wane. Much of that ill-advised insistence rests on the assumption that China's academic institutions will take a long time to become as capable as the best Western universities, perhaps as long as it took the United States to establish its own system of research universities. Yet China already has proved itself adept at siphoning off talent and expertise in much the same way the United States did from Europe following World War II. And let us not forget that China has lifted well more than 300 million people out of poverty in the last sixty years, nor that it has surpassed the Americas in the number of machine learning research papers in recent years.

---

* Kai-Fu Lee, "The Real Threat of Artificial Intelligence," *New York Times*, June 24, 2017.

The global competition in AI will raise as many new challenges as it solves; such is the currency of progress. But whatever pivots arise in this AI race, the applications that emerge from the global seedbed of innovation will change our world and benefit humanity in ways we can barely fathom today.

GEO-COGNITIVE POWER

Perhaps nothing underscores the deeper civil-military fusion enabled by AI and its cousin technologies better than their application in economic development programs. China has used its Belt and Road Initiative to encourage developing countries to play by its rules, essentially requiring that they bow to its notions of fairness, equity, and justice in exchange for the major investments it offers.* For example, when Kazakhstan joined about seventy other countries that have received Chinese-built infrastructure networks, it quickly saw a boom in traffic for the freshly upgraded ports along the highway and rail routes between the countries. In the first ten months of 2017, rail volumes at one station had doubled from the prior year.† Yet, those gifts came with strings attached. China's agreements favored its companies, products, and labor, making it more difficult for Kazakh officials to foster opportunities for their country's small and medium-size businesses. Months before some of the individual projects were implemented, Kazakh residents protested a provision that would've opened agricultural land to long-term leases by foreign companies.

Those countries might have to lean even further toward Beijing's sensibilities about the deployment and use of AI and related advanced technologies as well, especially given the fact that Chinese technology will ride atop of and be integrated into this new "smart infrastructure." That sort of *geo-cognitive power projection*—a technology-enabled race to spread and enhance a nation's global influence—could have far-reaching effects. China and other superpowers aren't merely trying to

---

* Nyshka Chandran, "China's plans for creating new international courts are raising fears of bias," CNBC, Feb. 1, 2018.

† Kemal Kirişci and Philippe Le Corre, *The new geopolitics of Central Asia: China vies for influence in Russia's backyard*, The Brookings Institution, Jan. 2, 2018.

project economic, political, and military power; they also want greater control over what people around the world believe, how they make decisions, and how much they let machines make decisions for them.

It might be hard to imagine about seventy countries with more than 60 percent of the world's population acquiescing to China's social credit system, but the people in those nations, who help produce about a third of global trade and about 40 percent of global GDP, might have to go along with those dictates if they aim to do business with the world's most populous market. Those governments and residents are coming to depend on the influx of Chinese mobile phones and the country's willingness to help developing markets build out multimillion-dollar networks, all based on technologies produced by Huawei and other Chinese titans. Artificial intelligence applications undoubtedly will ride on top of that development. In an era of waning US leadership around the world, much less a coherent industrialized-country vision for the global economy, China's vision might become an attractive alternative.

It's virtually impossible to assess all the second- and third-order ripple effects that the emergence of China as a potential geo-cognitive superpower could have on the global economic order, especially because it's not entirely clear what China will ultimately stand for. The unfolding of the post-World War II era with the American-led Bretton Woods Conference and the UN-style system followed more transparent and predictable lines. And while China is far older than the United States, its experience with Communist rule is just seventy years old—and it was evolving again with Xi Jinping's consolidation of power in early 2018. It is also not yet as experienced as the United States in designing and influencing global regimes that depend on it and draw countries to it. In addition, despite its large standing army, China has considerably less hard power to project via aircraft carriers and long-range bombers or cruise missiles.

The United States does have both the hard-projection capabilities and the soft power of its popular culture. It maintains a successful model of digitization among its population that has driven unparalleled economic progress. While that story still holds appeal in many places around the world, it has lost some of its luster, particularly in comparison to the emerging appeal of the Chinese narrative that is based on its own impressive success story. America's projection of AI applications, ethics, and influence still pervades the many countries in which its diplomatic and private-sector organizations operate. After all, Facebook would be the largest country in the world if subscribers were citizens. But in light of Cambridge Analytica and Facebook's other scandals, as well as the problems faced by America's other Digital Barons and its federal agencies, the extent of US influence might be in jeopardy.

6

Pandora's Box

No matter how careful, a perpetrator always leaves a footprint somewhere. To find it, Jeff Jonas looks in some unusual places. Jonas is not your typical data scientist. As of this writing, he was one of just four people who'd participated in every Ironman Triathlon on the world circuit. A system he developed for Las Vegas casinos helped catch the MIT blackjack team made famous in the best-selling book *Bringing Down the House* and the movie *21*. Even his latest venture, Senzing, was part of an unusual move by IBM, which agreed to spin out Jonas and his team into a separate firm that helps clients identify who's who across all their data sets. The process, called "entity resolution," helps companies prepare for the EU's new data-protection rules. But what he really, truly loves is finding that hidden footprint. "My particular passion is for systems that take down bad guys," Jonas says. "Helping our customers take down some real clever bastards brings me great joy."

He does this by uncovering unexpected connections across often-unrelated sets of information. (Among other things, he coded the idea into the Non-Obvious Relationship Awareness [NORA] software he created for casinos.) The approach looks across a range of data sets, including places so far out in left field that nefarious actors don't think about covering their tracks there. By piecing together relationships across multiple bodies of information, the process can start to develop a trail of evidence in the inconsistencies or discrepancies across various data sets. If you just moved into a new house and the guy across the street says he never travels overseas, you'd never know it's a lie. But then his wife gets drunk and says he lived in France. You just found an inconsistency. Simple enough in that case, but the challenge gets significantly tougher with criminals employing fancy tradecraft to try to cover their tracks.

Outside of someone making an obvious mistake or their accomplices giving them up, investigators have two options. First, they can search for the types of observations the perpetrators would never imagine or expect you to have—shifting into new observational space, as Jonas describes it. Second, investigators can use technology to compute in ways unknown to or unforeseen by the perpetrator. For example, an adversary might know you have a video camera, but he might not know anything about your license plate readers. Jonas's approach cuts across both of those. "You have to ask what data source might be available to present the contrarian evidence that would be more difficult for the bad actor to control," he explains. "What is the third record that might provide the glue to bind two other data points? And you might not know it's of interest until it acts as the glue."

Yet, for someone so deeply reliant on data to capture the bad guys, Jonas gets even more animated when talking about the need for individual privacy. He admits he didn't think much about it when trying to track card counters and other criminals in the Vegas casinos. But of late, he has made data privacy a concurrent pursuit alongside his company and his passion for finding bad guys. He despises the concept of the social credit score in China, calling it "possibly the most evil thing" because it "suppresses dissent and contrarian opinion." He thinks the EU's stringent data-protection regulations will likely become the standard for privacy regulations around the world. Yet, he has no problem with the inevitable growth of a surveillance society. "To surveil is to look. It's not bad. You surveil the street to see if it's safe to pass," he explains. "So, the primary point is *what* data is in your observation space, and do you have a legal and policy right to it."

One might debate, for example, whether Cambridge Analytica should've had a legal and ethical right to the data it scraped from Facebook for the sake of voter targeting, sparking a scandal about the social media platform's mishandling of members' private information. But the other side of it is that all too often, consumers jump right into the fray for irresistible product and service offers or for new digital experiences, willingly giving up their privacy every day for some benefit or convenience in return. The question isn't whether the AI systems that power the analyses of our lives are good or bad. They're tools that can be used for multiple simultaneous purposes—and they will continue to be, unless people are willing to read the fine print of the user agreement or opt out, a move that usually leaves them out of the flow of transactions and information.

For example, providing broad access to patient data might produce amazing breakthroughs for health care, facilitating discoveries at atomic and genetic levels that greatly improve human well-being for all of society. Yet, an individual might want to hold personal health information much more closely, raising tensions in turn about the results an AI system is then able to produce with the smaller set of data available to it. In these and so many other AI examples, the most critical questions center on the value-based decisions about where we draw boundaries between private and public rights. The same sort of distinction holds true in discussions of surveillance and privacy, Jonas says. Surveillance is the mere act of looking. The important concerns are who looks at what and the control citizens have over that decision. Those are questions of privacy and the value individuals and societies place on it. And while security in any country naturally requires surveillance, the extent of that surveillance and the boundaries of privacy can vary widely from one nation to the next.

Those lines often move, too. People might come to demand broader and more-intensive surveillance as powerful new threats arise, especially since we can safely assume many of today's costly advanced technologies will become cheaper and commonplace in the future. "What if common, everyday technology could enable a single person to kill 100 million people for $5?" Jonas asks. "What must happen then?" One of his friends, who works on privacy and civil liberties issues, once pushed further on that idea with another intriguing question: In this sort of future, are we better off keeping tighter control over who can access the ubiquitous surveillance, or opening access to the widest set of people? Power corrupts if unchecked by the crowd, his friend suggested, so perhaps we're better off making sure that ubiquitous surveillance is available to the masses rather than concentrating it. Given the global explosion of data and the existing open-source availability of so much AI code, the ecosystem's inclination already leans toward broader access. If that continues, as one would expect, the possibility of someone using cheaper and more powerful AI to disrupt our lives could heighten the need for ubiquitous surveillance in turn. It's an arms race of sorts, based on the vast exponential effects of the technology.

These already difficult challenges to our fundamental notions of values, trust, and power won't get any easier as the deeper integration of cognitive machines in our lives triggers new dilemmas and super-charges old ones. We need to consider these "North Star" questions as we help shape a future of beneficial AI. How do we respect and preserve privacy? How do we maintain human choice and agency? How can we ensure justice and fairness? How do we build technologies that enhance human creativity and empathy? Absent a consideration of these and other North Star issues, we leave ourselves ill-equipped to respond to the potentially harmful uses of AI-powered technologies and, just as importantly, ill-prepared to capitalize on the opportunity these powerful tools provide to build a prosperous future for humanity.

BLISSFUL IN THE FISHBOWL?

Closed-circuit camera systems have been the norm in UK public safety for decades. Originally intended to detect bomb-setting terrorists of the Irish Replication Army (IRA), which sought independence from English control in Northern Ireland, the surveillance systems have become a widespread tool for crime detection, especially in London and other larger cities. English authorities use facial recognition as a means of identifying persons of interest. Moscow employs a similar system, with about 160,000 installed cameras, though only a few thousand are active at any given time for cost reasons.* According to Artem Ermolaev, the

---

* James Vincent, "Moscow says its new facial recognition CCTV has already led to six arrests," *The Verge* (September 28, 2017).

spokesperson of the city's department of information technology, the system covers some 95 percent of the city's apartment buildings, has an identification-success ratio of approximately 30 percent, and led to six arrests in its initial trial period. Low light conditions are still tough on the cameras, but that will likely improve in the near term.

Manindra Majumdar has put camera systems and AI detection systems together in an effort to tackle two of India's most notorious problems in recent years: untended garbage and public harassment of women. After launching an image-based search and shopping app called GoFind, Majumdar created a new start-up called CityVision AI and submitted bids in two Indian cities to test surveillance systems that identify garbage dumping or potential crimes. (It since has submitted a bid for one surveillance program in Dubai and another in Toronto, where it relocated its base of operations.) In India, the CityVision system tracks areas where people commonly dump refuse, and then alerts authorities when bins need to be collected or informs them about patterns of illicit dumping activity in targeted areas. It also can be used to spot when minor harassment of women begins to escalate and pass along the relevant location data and video for police to review.

Because the Indian population is so much larger and more diverse than in most other countries, Majumdar says, people more readily accept surveillance as something necessary for safety. Privacy is a different issue, though, so CityVision takes steps to make sure it protects identities and leaves any direct enforcement action in the hands of the sanitation department or police.

It's not hard to imagine cities around the globe latching on to these types of technologies, especially as rising urbanization rates increase socioeconomic struggle and heighten the possibility of conflict. The need to manage and de-escalate tensions in crowded places likely will override the need for privacy, at least in public spaces. And as migration, trade, media consumption, and money flows across borders accelerate, governments will interconnect these systems to track individuals and assets moving from one jurisdiction to the next. One day, we might very well see this tracking ability as the hallmark of a safe and trustworthy society.

So, we move steadily toward a society of augmented alertness and awareness, one in which AI-powered systems see and can focus on what human eyes and brains are too limited to detect or process. They can detect subtle shifts in presence or patterns, while we are stuck within our narrow field of vision. So, the machine and its artificial awareness of our context could become man's best friend, like a dog able to sense an earthquake before it happens. We might in fact be seeing a new type of AI hive consciousness that enhances both our well-being and that of society.

Not long from now, an AI platform might combine traffic, weather, infrastructure, and other users' information into a guide far more comprehensive and convenient than Google Maps steering you around a traffic jam. A cloudy evening portends downpours the next morning. Runoff and infrastructure data suggest an 82 percent chance that the heavy rains will overload the sewer lines under repair along your typical route to work. You have several calls but no urgent in-person meetings on the calendar. So, the platform automatically adjusts your schedule to give you the option of staying home or taking an alternate way to the office. Google already aggregates much of this data in 2018; it wouldn't take a significant technological leap to combine those streams into this sort of analysis.

With access to more and more data and the imagination to integrate those streams in different ways, companies and developers would deliver even more convenience. The same system easily could spot a new conflict that arises on your spouse's calendar and adjust yours to pick up the kids and take them to soccer practice. It might capture the news of a likely airline strike in France and suggest you rebook your overnight flight from Paris to Frankfurt, where it can secure a train ticket that would allow you to see your goddaughter on her birthday. And it might adjust your nutrition and sleep plans for the transatlantic flight, ordering you heart-healthy meals and resetting your dose of melatonin to help maintain your blood pressure, cholesterol, and rest regimens.

None of this should happen without a user's control over settings for privacy and personal agency, whether for individual services or when agreeing to share data between friends or strangers. We already give up most data ownership to the government, the Digital Barons, and so many other service providers, usually without recourse or understanding of what they do with it. Efforts to secure personal data ownership and control have emerged around the world, and they likely will gain traction as abuses emerge and risks become more apparent. But even if we opt out, we face a different sort of hazard—sleepwalking into a surveilling AI that puts more trust into people with larger digital footprints. That rift will widen over the next decade as more people demand the right to opt in and out of AI-based services. We could see a bifurcation of populations into more and less participatory groups. That could lead to privileges for highly active users or those who abstain, creating deep new divides that will stir up legal and civil discord. Large portions of the global population would fall slowly but steadily behind, as advantaged markets and their AI systems overlook the needs of the less fortunate. Politicians will see social stability, economic growth, and voter confidence waver. Our institutions will face an entirely new breed of inequality.

Being predictable and calculable will generate power, while the Luddites of the digital era might need to think hard and fast about entering urban areas—or create digital cloaks or twins to avoid detection. Will we have a moral right to remain ambiguous and unassessed, and how many of us would even want that? It's not hard to imagine more people opting out of certain AI-powered platforms to avoid what they see as malicious use. Facebook and other social media sites regularly face defections, some more serious than others. It's also easy to imagine technological countermeasures people might adopt to avoid attacks or unwarranted use of personal data.

Ultimately, like any new technology, AI provides us with another double-edged sword. The distributed ledger of an established blockchain platform makes transactions more secure and extremely difficult to forge, but it also allows users to remain anonymous and hinders law-enforcement efforts to track criminal activity. Similarly, advances in AI technologies might one day allow school officials to identify a troubled student and intervene before they bring a gun to school, but the same innovations could help students hide from accountability and responsibility.

THE BOUNCER BOT

New applications will emerge to safeguard our privacy, make us less transparent and readable, and protect our digital personas. Some of these ventures exist already. "Controlio, for instance, acts as an intermediary between you and the larger internet platforms, issuing RFPs for products or services on your behalf," says Peter Schwartz, founder of scenario planning consultancy Global Business Network and now Salesforce's resident chief futurist. "When you transact, you still give up data, but you're in charge on a case-by-case basis."

As AI evolves, we could see a whole new layer of a personalized-data economy, including personal-data vaults that open only when we want them to, rather than the constant data giveaway model of today as soon as you step into the web. However, we can't optimize for choice and security simultaneously; we need to strike a balance, one that shifts depending on context—giving mom and dad access to photos of the grandkids, but barring social media sites from manipulating data about those children. Companies and governments already try *digital strong-arming*, wrapping their wants together with the products and services we crave to pull more out of us than we'd prefer to share. So, perhaps we'll create new protective agents, what we might call "Bouncer Bots."

Bouncer Bots would patrol the velvet ropes around our data, allowing immediate access to those we want, holding off some requests for a more-thorough review, and rejecting certain others. They could act on our behalf, looking for what we desire, offering short bursts of data or virtual currencies in exchange for those things, and then reraising the wall of personal protection. The Internet giants and most companies would reject the concept, perhaps even refusing services to consumers who bring their own Bouncer Bots along to their platforms. From their perspective, that's a valid objection. After all, for the better part of twenty years we have benefited from services that were deemed "free of charge" to users. Lately, however, the concept of "free" has morphed into a clearer understanding of what users pay when they turn over their data, and AI will drive that further.

But past examples, such as the radical transformation of the music industry by digital technologies, have taught us that fighting customers instead of giving them what they want never works for long. The virtual standoff between data-driven companies and individual consumer privacy will settle into some sort of parity with a workable business model. Still, we find regular instances of these technologies crossing the lines of what we find acceptable. We rarely object, but, when we do, the transgression often seems outrageous, like Mattel's efforts to help learn from and improve a child's experience with an advanced version of its iconic Barbie doll. Since withdrawn from the market, the enhanced doll collected and stored information about how children played, responded, and spoke with their Barbie, shipping all that information back to Mattel's servers. From the data, the company could glean insights about behavior and development while offering other targeted services, such as child monitoring for parents. The idea that Mattel might track children and use that data to target products at them chilled many parents, especially as a commercial interest reaping the data of an unknowing minor. Parents might share a lot of data about themselves and their children; they don't want Barbie collecting it.

Yet, they already supply that sort of information in ways that are far more intimate. Every minute of every day, smartphones collect our locations and behaviors. With the right accessories, they even monitor our sleeping patterns. Companies can use those data streams to provide an array of services. Already, Apple and Android phones identify when you stopped driving and started walking, so it can remind you where you parked your car. Or they can identify that you're driving and block incoming texts until you stop. Yet, smartphones can gather much more granular data to generate powerful insights, including into someone's medical state, a use that has allowed one company to push mental-health services and monitoring into participating patients' everyday lives.

Serious mental illness produces a cyclical pattern of inpatient care, release, relapse, and re-admittance. About a third of patients treated for mental illness return within a year. It's a vicious cycle that few technologies or treatment methods have managed to break in any comprehensive manner. Paul Dagum and his team at Mindstrong hope to change that. Mindstrong uses extremely fine measurements of activity on cell phones and certain other devices to track patient conditions. All told, Dagum says, they generate about 1,000 markers of cognitive capability from things as straightforward as millisecond-response times or patterns of finger flicks across a smartphone.

Machine learning helps compile and analyze the fine-grained patterns, which can show when a patient's cognitive capabilities start to weaken and alert caregivers or family members who can intervene before the deterioration goes too far. The patient and care-team apps allow the two sides to interact and look at cognitive markers together, identifying potential triggers, recalibrating medications, or just stepping up outpatient therapy. "Now, we're mostly focused on patients with a mental disorder or at high risk for developing one," Dagum says. "The response is largely positive because people feel vulnerable, and this gives them a sense of comfort. But that only works because we approach them as health care providers within the health care system."

Eventually, the cognitive monitoring on the Mindstrong platform could expand to track dementing illnesses, such as Alzheimer's, or a host of other ailments that affect cognitive function. Eventually, Dagum hopes, mental health care will become part of everyone's general health maintenance. Pharmaceutical companies already have approached Mindstrong to use its markers to generate more insight into drug testing. But for now, it starts with serious mental health issues and meeting them where they live their everyday lives. Dagum says he expects the technology to cut the readmission rate for patients by half. "This moves care out into the community," he says. "That will significantly affect the outcome for these patients."

We still have to ask ourselves what AI platforms and their owners can measure with our hundreds of keyboard strokes, mouse clicks, and smartphone swipes each day. The technology research firm IDC estimates that people connected to the Internet will increase their average digital interactions from 218 a day in 2015 to almost 4,800 a day by 2025.* Will each of these be subject to psychoanalysis? Many could be, as new medical and well-being applications proliferate. And as the processors and sensors in our mobile phones become increasingly sophisticated, more and more physiological data can be cross-referenced and correlated with digital behaviors. With something so intimate as our physical and mental health, we will want to ensure that such a system can make users aware of the data it's mining and how it will use that information. We might require that service providers alert users, their families, primary care physicians, or public health officials when new illnesses present themselves, especially when those ailments pose a risk to others. As those choice moments occur, thinking machines can improve health and well-being on both an individual and community level. But in so doing, they will force

---

* David Reinsel, John Gantz, and John Rydning, "Data Age 2025: The Evolution of Data to Life-Critical," IDC white paper sponsored by Seagate Technology (April 2017).

us to make difficult decisions about the balance between individual privacy and public safety.

No doubt, there will be breaches of confidence as companies and AI platforms collect too much information and pressure patients and providers with new burdens of responsibility. Paranoia might arise as users see more of their biometric data harvested each day by their smartphones, watches, and other devices but don't immediately realize that the benefits of that information-sharing might not come for years. Medical AI providers will harvest the riches of data troves for drug recommendations and advertisements, and some might try to push the regulatory and ethical boundaries established by the Food and Drug Administration or other entities. Those wounds will cut deeply, because medical information is harder to recover than financial information, but we have a base of prior experience in enforcement of data portability, privacy, and insurance regulations. And, as with Facebook and Twitter today, public awareness will rise, popular backlash will increase, and enforcement agencies will start to address the violations.

CONSTANT BECOMING OR DIGITAL REWINDING?

As human beings we are notoriously biased, too often unaware of our intellectual and emotional blind spots. A well-crafted artificial intelligence, even with its own shortcomings, could help us make richer, more-objective decisions that improve our lives and communities. Such a system might provide an alternative option for your work commute, offering a plan that balances a sharp reduction in your carbon footprint with enough convenience that you don't quickly abandon the new travel plan. A look at the divorce rates in most industrialized countries might lead one to believe that a few objective, analytical pointers about partner selection might not hurt. Teachers could use thinking machines to craft more effective curricula tailored for students with different learning profiles that update in real time. American AI experts already are working on systems that can help us avoid food shortages and famines by integrating changes in factors like weather, soil, infrastructure, and markets into complex models to mitigate scarcity.

Beaconforce may not be solving world hunger, but they've found a way to help alleviate something most of us deal with on a regular basis—the types of workplace stress that keep us from performing at our best. The company's system tracks clients' workers along seven pillars that contribute to "intrinsic motivation," the type of drive we feel when we're immersed in an engaging and rewarding activity. These pillars, which include feedback, social interaction, and a sense of control, feed into a worker's ability to stay "in the flow," as CEO Luca Rosetti describes it. When the balance of a worker's abilities and the challenges they face tips too far, the Beaconforce dashboard can alert managers.

It does this with an AI-powered analysis of worker sentiment and certain vital signs measured by a Fitbit, Apple watch, or similar wearable device. The program asks a client's employees a couple quick questions each day on their smartphone, and then correlates those answers with information about their current work environment and their heart rate. A manager can't see their workers' individual answers or heart rate, but they can see the Beaconforce dashboard, which signals when a worker is starting to feel out of sorts about a project, coworker, or environment.

Rosetti shared four testimonials, including a story from a partner at one of the Big Four accounting and consulting firms. The partner at the company noticed that three of his employees had shifted suddenly into the stress range of the dashboard Beaconforce provides. He didn't pay much attention at first, because moments of stress are commonplace in their line of work. But then a human resources officer came in and said one of the consultants had an anxiety attack on that day, breaking down and crying in a meeting but refusing to say anything about why. The manager immediately guessed who it was and started to investigate the issue.

The Beaconforce platform showed when all three workers' readings initially started to deteriorate, and it aligned with their assignment to the same project leader. It turned out the project leader was consulting for another company and had pressured all three to join him, threatening to make life miserable on their current project if they refused. So, the partner swapped out the project leader and immediately saw the workers' scores recover. The partner even managed to retain the project leader, who turned out to be extremely talented, according to the case study Rosetti provided.

That sort of AI deployment can help facilitate greater achievement if we design it well, but that blade cuts both ways. Cognitive computing can increase or decrease our freedom of choice, but the risk of the latter increases with the large-scale collection and manipulation of personal data. We risk tipping the parity between what we know and what the machine knows about us. Artificial intelligence might enhance our abilities, but without a basic parity of awareness between individuals and the entities that control our data it might also limit our fullest potential as we sacrifice our own self-determination. Similarly, replacing human-curated judgment with machine-curated judgment might broaden or narrow our field of vision, and it might reduce or expand our social and economic choices—often without our knowing which way and by how much. Taken individually, the nudges of mercantile and political interests might have little consequence. Collectively, they can transform our lives in powerful ways, like Cambridge Analytica's deployment of targeted messages to sway millions of US voters in 2016.

The mere push toward a more equitable balance of awareness will help expose many of the hard-to-define tipping points between the beneficial and manipulative uses of AI systems. This effort should begin with a reset of data transparency and control, allowing each person access to the information collected on them and the ability to expunge it or port it to, say, a new job or health care provider. We might develop a new structure for opt-in agreements that includes temporary opt-out rights, giving users a chance to step back out of the bubble. Workers might have an option to pause productivity nudges, for example, or someone with high cholesterol might stop their alerts and enjoy a nice steak on their birthday. Measurements and data insights, in and of themselves, might not manipulate, abuse, or intrude. But if we ultimately hope to empower each person to live a more potent, productive, and fulfilling life, they must retain the agency to decide how much is enough, without the possible alienation that might come with opting out of a service.

Ultimately, though, identifying and monitoring that thin and blurry line between benefit and harm will require greater transparency and public scrutiny of new algorithms, data sets, and platforms. Despite the open atmosphere across much of the AI field, proprietary code and data often make it difficult to ascertain the point at which systems cross boundaries. That uncertainty will lead to problems in legal proceedings, financial transactions, and virtually every other field that AI touches, but the concerns it raises for education are especially critical. We can barely agree on the line between education and brainwashing in our analog textbooks and longstanding curricula, let alone in areas where AI systems provide far more subtle nudges toward learning outcomes.

By keeping a human in the loop as part of the peer-to-peer tutoring platform he's developing, Patrick Poirier hopes to install a crucial guardrail against AI manipulation. Poirier founded Erudite AI, a Montreal start-up that uses machine learning to increase the effectiveness of students tutoring other students. The application, still in the pilot stages in early 2018, will help guide students as they work with a fellow classmate, analyzing their interaction, integrating a variety of best learning practices, and guiding the tutor toward the best educational pathways for the student with whom they're working. For example, if the tutoring student tries to provide an answer immediately, the system would interject and help the tutor provide feedback in a way that helps his fellow pupil learn, Poirier explains. As he and his colleagues train the system, a professional educator will monitor the interactions. At the outset, Erudite needs the trained teacher as a proxy for the AI system they're developing. Those interactions begin to prime the AI, and then educators can help refine its capabilities as it takes on more of the interaction. Ultimately, Poirier says, the largely automated system will guide peers as they help each other learn, but it also will remain auditable by teachers or educational agencies.

That initial idea drew the attention of the IBM Watson XPRIZE officials, who named it a Top Ten team, but what makes the start-up especially unique is Poirier's broader outlook on the future of education. Rather than grading school exams down from 100 percent, the "disheartening" current standard, Poirier says, Erudite will work on a "grading up" scale, much like a video game might. Students start at a baseline and work their way higher as they gain more knowledge. "With technology, we can see more personalized learning, more interactive learning," Poirier says. "The theory will be integrated into the act of learning." Whereas students today come to professors wondering where they lost points—as if they'd ever earned them in the first place—they will have to work up to those scores in the future.

"You can learn by doing tasks, but if something isn't validating that you're learning the task, you don't know. You need to test to measure that," he says. "With technology, you can start grading up because as you do things you get points."

Such approaches won't guarantee the validity and veracity of AI systems, but resetting goals to incentivize gains over penalties could help enhance trust in thinking machines until fuller transparency and "explainability" emerges. In the meantime, evolving AI technologies will improve our ability to analyze algorithms and data sets. Already, researchers have developed AI models that test one another, pushing toward higher quality output (at least as defined by the developers behind the code). Only public scrutiny can assure the quality of those algorithms and data sets from a broader societal perspective. Continued efforts to establish bodies that can audit AI systems and balance corporate interests with community values could help safeguard against the manipulation of learning minds.

SAVANT MACHINES OR SYMBIOTIC COGNITION?

Companies collect, analyze, and sell massive streams of data on each of us. Walk into the grocery store, and security cameras track the route you take around the aisles. Swipe your credit and loyalty club cards, and the store compares what you purchased and in what combination with your previous trips. Pull that all together, and the store knows precisely what product to put where and at what price to make you (and other similar shoppers) more likely to grab an extra pack or two. Yet, even the grocery store, let alone a huge Internet company, has far more data about you than it would ever care to use. Amazon might not need to know a customer is a physically fit, middle-aged, cheese-eating, blues-loving, job-hopping, Latino fly fisherman with ancestral roots in Costa Rica—unless that combination of attributes make them more likely to buy a certain product.

The depth of awareness that large companies could compile differs from what they actually compile, and that might help quell some of our anxieties. Much like the blurry lines of confidentiality and privacy, we don't have a clear sense of where the balance tips on the equality of power between the system and the person. If companies begin to generate ever-finer grains of truth about our life patterns, do we really want to see them in their raw form? Does that knowledge begin to undercut the tenuous sense of parity that relies, in part, on the willful ignorance of consumers? After all, under the EU's data-protection rules, people retain control over the primary data companies collect on them, but the companies appear to maintain control over the insights they derive from that data. We might not really know how they judge us.

In 2017, we visited the IBM Watson West center with a delegation of executive MBA students from the creative industries, during which the IBMers there demonstrated a new application that analyzes personalities based on Twitter feeds. One of the students, an outgoing studio-musician-turned-entrepreneur from Texas named John Anthony Martinez, bravely volunteered his feed. Watson promptly concluded that Martinez was "melancholy" and "guarded and particular," but also "energetic" with a fast-paced and busy schedule. And while Martinez preferred to listen rather than talk, Watson concluded, he's "driven by a desire for self-expression." The results didn't surprise Martinez, but he didn't fully agree, either, perhaps because Watson doesn't discriminate between tweets from retweets, which probably skewed the report a bit. People with more restrained sensibilities might object to many of those labels, especially when based on a data stream that captures just one slice of their identity (and one that's often fabricated, to boot). Sometimes, that unvarnished feedback provides valuable insight, but experienced coaches and psychologists say that works only if the feedback is delivered properly. Could Watson gain enough emotional intelligence to share its awareness fruitfully, in a way that maintains a healthy, equitable relationship between coach and individual, machine and human?

It's an entirely different matter when it's another person, not a machine, making such judgments. Plenty of people are inept when it comes to emotional intelligence, empathy, and awareness of emotionally dicey situations. But we know that and, absent evidence of emotional cruelty or intent to injure or slander, we accept a fair amount of ambiguity in our human interactions. We know and trust that humans, in general, possess the sensory and evaluative capabilities to pull back from or ease explosive situations.

We don't yet know if that sensibility will exist in cognitive computers. How do we assure awareness parity for especially sensitive social questions, such as sexuality, identity, and personal health? Most of us might readily accept the medical advances powered by IBM Watson if they can provide doctors a clearer picture of our well-being, even if that means the machine knows aspects of our minds and bodies better than we do ourselves. And we might want an AI-powered system to provide us sound financial advice and nudge us toward better use of our hard-earned money. But if they do know us that intimately, don't we want to know how deep that insight goes?

Imagine the machine developing a constantly evolving analysis of you and the people with whom you relate, but never communicating that view back to you or them. Years from now, an AI system might help "George" prepare for the day ahead by collecting personality profiles on all the people he expects to meet. The platform wouldn't need access to the various individuals' data streams, having already scraped enough information from publicly available social media and its analysis of any relevant voice recordings, personal chats, or photos on George's smartphone. Now, imagine you're sitting across the table from George, getting his feedback about your ongoing job search in the United States. While you're asking George for advice, his AI is processing the discussion in real time. It's analyzing your chances of landing a US-based job and correlating that with data gleaned from other workers in related industries. Does George use that feedback to mentor and advise you? Or does he realize your prospects are dim and dismiss you because you offer little benefit as a member of his business network? Meanwhile, having come to the United States from Europe, your AI assistant might be bound by the EU's stricter privacy protection laws, limiting the depth of the analysis you receive and leaving you with far less insight about George and his motivations.

Cognition is power, and a range of individual and societal factors will influence how we balance that power. Different regulatory regimes, willingness to participate in the digital economy, and the ability to afford better products will affect the unequal distribution of cognitive power. None of these imbalances will be easy to correct—and, as a society, we might decide some should remain—but we could start by taking a global view of AI and the flow of personal data. Perhaps fifteen to twenty years from now, we might sign an accord to limit asymmetric awareness in AIs designed or hosted in another region, starting an era of cognitive balance in which the power of knowledge is constantly renegotiated, tested, violated, verified, and reset.

TRUTH AND TRANSPARENCY FOR TRUST

Human values and trust each rely on a certain understanding between people, one that requires a minimum threshold of trust. Those levels can vary, of course. We might not need to know the mechanic who fixes our car, and we probably don't worry about whether he signed an oath to uphold the best auto maintenance practices, but we certainly expect both from the surgeon about to operate on us. Regardless of whether we want to know precisely what the doctor will do as she digs around our innards, we need the assurance that she's subject to a higher requirement for transparency and explicability. Absent that, we lose trust in the process and the person.

To maintain similar levels of trust and shared values—and, by extension, to retain an appropriate balance of human-machine power—the increasingly pervasive role of AI will raise the threshold we demand of transparency and "explainability," as it's referred to by developers. The use of AI-powered policing in Los Angeles and probation systems in US courts has raised objections from constituents who demand a right to know how they are being ranked, rated, and assessed and what the logic was in the assessment. Yet, as Wade Shen noted in Chapter 4, finding a way to make an AI system explain the how and why of its decision is a difficult challenge, one in which DARPA has invested millions of dollars. The project has significant ramifications for military leaders who seek to understand why a system recommends an action that will have deadly consequences. But it also plays out in day to day life, for example as federal investigators and Uber's scientists try to figure out why an autonomous car hit a pedestrian in Phoenix. As self-driving vehicles become more commonplace on streets around the globe, will the systems that operate them be able to explain why they made decisions that led to damage or death? If they can't, would we ever have enough trust to remove the steering wheel entirely?

Given its June 2018 effective date, the EU's data-protection directive looked much further into the future than most existing AI-related regulation. Its one-size-fits-all approach ruffled feathers, as many applications don't need perfect explanations. In fact, society might prefer less-than-perfect explanations at times—say, if a sophisticated medical diagnostic AI works better than any other system to identify an illness, but is so complex we don't fully understand why it's so effective. We might want it to enhance our well-being, perhaps even save lives, but we can't be certain of the precise reasoning that gets us there. Legitimate questions about how much trust we place in that system, how sound it is over the long term, and its unintended consequences might become moot, because the regulations threaten to stunt it from the start.

Yet, Europe also stepped up as a white knight against the massive market power of the Microsoft, Google, and others. While the region might not produce one of its own Digital Barons given its regulatory leanings, it has started to unify a single digital market with 500 million people and has set out plans that could produce a sort of "digital-commons baron." The EU's latest goal to create a collection of large, public, and open data sets for training AI systems could provide the fuel for a vast array of new models and applications, including many we have yet to even imagine. Even the most AI-savvy companies have more data than they know what to do with, much of it irrelevant to their business mode. (In 2018, one online service was asking prospective hires how they might think about putting its superfluous data to good use.)

Creating and controlling those widely available data sets will only expand the EU's existing influence on transatlantic mindsets and business models, and many observers expect markets around the world to adopt forms of its data-protection regulations. Furthermore, without the transparency and explainability at the heart of the EU rules, we have a much more difficult task ascertaining a system's veracity. We can gauge the outputs and measure how they change with different inputs, essentially distilling a better sense for the integrity of the final results. Yet, one wonders how policy makers and regulators might seek to put boundaries on processes that can only be measured in hindsight, especially when those systems influence our health, guide the distribution of government resources, or conflict with societal and cultural values. For applications that don't have a direct influence on people's lives, a retrospective explanation of the machine's decision-making process might be enough. We might not care at all about some things—how much does it matter if an AI-powered guidance system tells us to take Park Avenue instead of Madison?—but others may get under our skin in a deeper way.

As these applications play a deeper role in individual lives, they require deeper levels of trust, and thus confront us with a series of trade-offs, a continuum along which benefits are measured against ignorance. In 2018, millions of people put their trust in online dating sites, trusting the algorithms of Match.com and others to find compatible mates. They trust the process, in part because it's not any more error-prone than the old-fashioned method of randomly meeting someone and getting to know them better. Fast forward a couple decades, when matchmaking systems have more data and deeper training in human compatibility, and we might want a little more explanation about how the systems make those matches—or, perhaps more realistically, why it didn't make a certain connection.

Perhaps by then our AI assistants will pick up on imperceptible cues and data during a first date and identify it as an inevitably flawed pairing. He might display subtle physical cues of compulsive lying; she might have left faint data trails of an unfaithful past. Yet, the date went well, and they felt like it scored on all the personal, physical, and emotional attractions they hoped for. If they decide to ignore the machine, they might head down a path doomed to heartbreak. If they follow its advice and decline the second date, they might miss out on a serendipitous connection that defies the odds. Absent a clear explanation of why the machine questions the connection, how do they trust it to make that call? And what if the company's AI platform begins to identify all their frailties and decides they're too risky to match with anyone? Deep learning algorithms might consider matchmaking in a neutral, transactional activity—a statistically determined gateway to a first date, rather than a headlong plunge into a relationship with all its potential for self-improvement and growth.

Of course, nothing is certain in life. Some relationships work, some don't. Even machine learning and neural networks generate probabilities based on past patterns, not perfect predictability. So, we might shrug off a first date that doesn't work out. However, we will demand greater insight when, say, military or political leaders make decisions based on an AI analysis of intelligence. A soldier called to war might reasonably question why political and military leaders made the decision to put them in the line of fire. A soldier's family and a nation's citizenry might reasonably demand better explanations and less ambiguity about how and why the

AI recommended lethal and dangerous action. (This might be especially true in the United States, after faulty intelligence about weapons of mass destruction launched the second American invasion of Iraq.)

As we approach greater risks to our emotional and physical well-being, we heighten the levels of trust we require in what we can't understand. The same will hold true in an era of thinking machines. Until AI systems can explain their decisions well enough to foster a deeper sense of trust, people will naturally err on the side of caution and limit the benefits artificial intelligence could deliver. Over time, we will experiment our way toward explainability, feeling out our comfort level with the lack of certainty in one application or another. In some cases, we will take comfort from other situations in which we don't care about perfect clarity, such has traffic routings. But we also will double down on efforts to enhance our understanding of trust ratings, medical diagnoses, and the many other areas where explainability is essential.

DAWN OF A NEW DEMOCRACY OR DIGITAL FEUDALISM?

Because they rely on a constant diet of data, thinking machines threaten to expand the digital divide between the connected and disconnected. People in developing countries, especially in rural areas, will reap fewer of the benefits offered by AI-powered systems. Those who can afford greater access or deeper interactions stand to build on their advantage, as those who produce an economically valuable footprint are granted the option to barter their data for ever-expanding access. At worst, they at least have the choice to opt out. Even in affluent, data-rich places, artificial intelligence could polarize societies by pushing people further into bubbles of like-mindedness, reinforcing beliefs and values and limiting the sorts of interactions and friction that force a deeper consideration of those principles, something we in American societies have already witnessed. Taken a step further, AI systems could facilitate digital social engineering, creating parallel microsocieties, allowing the broader ability of companies to target job seekers of certain demographics and exclude others, as we've already seen on Facebook.*

---

* Julia Angwin, Noam Scheiber and Ariana Tobin, "Facebook Job Ads Raise Concerns About Age Discrimination," *New York Times*, Dec. 20, 2017.

Increasingly, we will run into situations where the data generated in the off-line sphere of our lives will not be counted by those who primarily focus on the types of digital information they can readily and efficiently process into "actionable insights." That's just fine for those who don't want to be captured, but it limits their ability to determine their standing in groups, communities, and societies. This evolving balance of power between *data serfs* and *data lords* hints at a new digital feudalism, in which those who provide the least digital value find themselves left with the fewest options. It's a transaction that favors the willing customer, but especially the owners and designers of the platforms, including the Digital Barons.

The feudal metaphor extends to the workplace. Companies that drive these relationships often enjoy a less costly and more flexible workforce. Yet, it also goes even deeper, to a more fundamental shift in the labor and consumer economies. The concept of a job as one coherent employment relationship to which one brings a specialized set of skills is being disaggregated, with the pieces offered to the awaiting online crowd, which picks over the them and fits them into their own lifestyles or financial pictures. This works well for the well-educated or the young, who have the in-demand skills companies need or the time and resources to accommodate changes in demand for their talents. It works less well for those who need predictability to support their family's minimum viable livelihood.

The ironic part is that the data lords often appear much like the Wizard of Oz, powerful in large part because of their ability to pull the levers and push the buttons as "the man behind the curtain." We see it in the varied approaches to autonomous cars. Many of today's so-called self-driving cars and trucks have remote drivers at the ready, sitting in control cockpits and ready to take over when a vehicle encounters a novel situation, such as inclement weather, debris on the road, or a construction site with a flag-man waving you over to drive between the cones on the wrong side of the highway. This is the strategy Roadstar.ai employs for its robo-taxis in Shenzhen, and that Los Angeles-based Starsky Robotics uses for its robo-trucks.

It's not just the self-driving cars that use the man behind the curtain, either. Data labeling and remote human participation help train many AI systems today. Cloud services that perform visual recognition have long had humans in the loop to identify the images and videos the machine labels incorrectly, in some cases employing thousands of workers overseas. This gives customers better service while building up the already-huge corpus of human-labeled data. In the case of medical imaging, providers must rely on the "tool" model, where the system merely suggests interpretations or notes areas of potential interest in images, the diagnosis left up to human radiologists. Education systems can incorporate a human touch in the interaction by letting a real teacher listen in on a lesson and change the responses in real time, much like Erudite AI does in its current iteration. This is used later to train the system to give the same responses in similar situations and reduce the need for human participation, potentially freeing teachers to engage in even richer interactions with students. In such specific settings, researchers believe realistic, humanlike interaction is achievable, but only in narrow forms.

Like the Wizard of Oz himself, developers might ask us to pay no attention to the man behind the curtain, but the concept of a human in the loop will remain indispensable for decades to come for certain systems, especially when it comes to ethical and moral decisions. The US Department of Defense still requires a human to make decisions on the use of lethal force, typically thinking about automated processes in terms of decision trees, says Matt Hummer, the director of analytics and advisory services at Govini. The question, then, is where on that decision tree does the human reside? Does a human pull the trigger or press the shiny red button, or does that person merely monitor the actions of the system to ensure proper function? Hummer can imagine times when the military could rely on machines to make automated decisions involving lethal force, particularly in defensive scenarios when time is critical. The Department of Defense has invested heavily in virtual reality and other battle simulation systems to help train systems, Hummer says, but most believe a human will remain in the loop and AI will "do a lot of training that will create those decision trees for us in the future."

Humans recognize a significant difference between a mission-driven machine making a fatal mistake and a person making a similarly faulty decision. We allow for errors when a person makes a decision—nobody's perfect, after all—but we expect the machines to work right every time. What happens when innocents are illegally killed? Military leaders can't court-martial a machine, especially one that can't explain how it made its decision. Do they prosecute the developer, or the monitor, or the soldiers who called in the AI-driven system from the field? And how do they ensure that the consequences of military applications remain confined to the battlefield? That's where the training needs to be impeccable, says Hummer, with safeguards in place if the machine encounters an uncertain situation. For example, an AI-assisted weapon should recognize a nonmilitary situation and refuse to fire. "But even then, we can have faulty AI and have a bad situation," he says.

Military use pushes these questions to an extreme edge, albeit an important one to consider given the billions of dollars flowing into AI-powered defense applications around the world. To address the common-yet-critical ethical scenarios people might encounter on an everyday basis, Peter Haas and his colleagues at Brown University's Humanity Centered Robotics Initiative have taken a novel approach to the human in the loop concept. One piece of their cross-disciplinary approach puts a human and a machine together in a virtual reality simulation, allowing the machine to learn from the human as he or she makes decisions and performs actions. The system works for basic understanding and manipulation, but Haas, the associate director of the center, says the strategy also works in a broader initiative to tie morals to scenes and objects.

He explains: "In a specific scene, a certain set of objects determines a certain behavioral pattern. So, you see a scene and it has desks, a blackboard, and there's a bunch of children sitting at the desks. You're going to expect this is some sort of school scenario. We're trying to understand, if that's the expectation, are there certain objects that would change the expectations of the scene? If you see a gun in that scene, then you see danger or a security guard or something like that. You're looking for other objects to figure out what the expectation for behavior is."

Currently, they perform the research in a virtual reality setting. The goal is to prepare more capable robots for interaction within societal environments, basing system norms on human behaviors and ensuring their relevance to the objects and context of their surroundings, Haas says. It doesn't take too active an imagination to see how various AI systems might come together in that situation. In the schoolroom scenario, for example, facial recognition might kick in to instantly identify the gun-toting person as a police officer who's visiting the class. "The advantage that we have for robots and AI agents is that we have the ability to draw on large databases of information that humans might not have access to immediately," Haas says. A human security officer could simply not do it as quickly and error-free in a live threat situation. "Robots and AI agents can quickly leverage big data to solve problems that humans couldn't, but AI agents don't have any of the moral competency that humans possess."

They're even further from understanding the variation of norms from one situation to another, let alone one culture to the next. One of Haas's colleagues has researched cultural norms in Japan and the United States to identify commonalities around which they can begin developing moralistic behaviors in AI agents. Over the coming decade or two, he and his colleagues imagine widespread global participation, with augmented and virtual reality systems helping gather human reactions and behaviors that can help create a moral and ethical library of actions and reactions from which robots and other AI systems can draw.

What makes the center's approach so intriguing is its aim to incorporate as wide a set of human inputs as possible. AI developers will have to put many of our grayest ethical debates into black-and-white code. Whose norms do they choose—and are those norms representative of human diversity? Algorithms have expanded beyond mere tools to optimize processing power in computer chips, match advertisements with receptive audiences, or find a closely matched romantic partner. Now, code influences far less obvious decisions with far more ambiguous trade-offs, including our voting preferences, whether we want our car to swerve off the road to avoid a dog, or whether a visitor in a classroom presents a threat to our children.

A DIRECT LINK INTO PHYSICAL HEARTS AND MINDS

These examples already raise critical questions about digital democracy or digital feudalism, but what happens when the machine connects directly into the human brain and/or body? How do we ensure trust and a balance of power when direct computer-brain interfaces could understand neural processes, read intentions, and stimulate neurons that enable hearing, vision, tactile sensations, and movement? Phillip Alvelda remembers the first time a DARPA-supported haptics technology allowed an injured man to feel an object with a prosthetic arm and fingers. The guy joked about becoming right-handed again, and the crowd laughed through the tears, says Alvelda, a former program manager at DARPA's Biological Technologies Office. Now, scientists can induce feelings of touch, pressure, pain, and about 250 other sensations. "The medical treatments are legion," he says now, speaking in his post-DARPA capacity. "We can build artificial systems to replace and compensate for parts of the brain that were damaged."

These cortical chip implants already can help alleviate tremors in Parkinson's patients, restore sight, and help overcome damage from strokes or other brain traumas, and researchers already have the fundamental understanding necessary to dramatically expand their capabilities. Neuroscientists already can identify the brain's abstraction of various ideas—for example, they can track the regions that light up in response to very specific concepts encoded in language, Alvelda explains. "If I can understand the core piece, we don't need the language," he says. "We can interface not at the level of words, but at the level of concepts and ideas. There's no reason it should be limited to seeing or sensing. There's no reason we couldn't communicate at the level of feelings or emotions."

At the South by Southwest Festival in March 2017, Bryan Johnson pointed out just how inefficient it was to have hundreds of people sitting quietly in the same room and listening to just three or four panelists for an hour. Johnson, the founder of Kernel, a company working on brain implant technologies, marveled at the aggregate brainpower gathered in the hotel ballroom that afternoon. "What if we could have a conversation between all of us at the same time?" he asked. That sort of hive communication, supplemented by an AI to help process signal from noise, might allow a crowd to instantly share and process all their collective emotions, concepts, and knowledge, perhaps gaining efficiency by activating rich neurological processes or communicating in images instead words.

That depth of communication, if even possible, remains many technological breakthroughs away from feasibility. For example, such a system would have to cut through enough noise to ensure human brains can process vast streams of incoming information in a room with lots of participants. And given our reliance on so many visual, tonal, and gestural cues to enhance our communication and understanding, we might lose meaning in the interactions that don't include those signals. But this is not the stuff of fantastical science fiction anymore; researchers have already created neural implants that supplement or replace lost brain function. "With the progress we've made in recent years, these are things we could start building now," Alvelda says. "This is not a speculative thing. We know which part of the brain does it. We're beginning to understand the coding. We can implant the devices." It will take time to get regulatory approvals for brain surgeries, so the initial cases will involve injured patients with few other options. And some technical challenges remain, including work to expand the bandwidth of data feeds in and out of the chip, but researchers don't need a grand breakthrough to make all this work, Alvelda says. "Today we're making devices that can perform the functions for animals, and we're a year or two at most from doing these experiments in humans with full bandwidth and wireless connection," he says. The technology and procedure will still need to go through the FDA protocol, but the timeline won't be measured in decades. "The [time] between the moment when we write the first synthetic messaging into a human, to the point where we get it commercial—where a blind person has an artificial vision system implanted in the skull—that's seven or eight years from where we are today," in 2018.

That's not a lot of time to figure out the thorny ethical problems that arise with direct links to the human brain. The mere idea of neural implants raises enough red flags, concerns that Alvelda and his colleagues, peers, and predecessors at DARPA readily acknowledge. On the most basic level, there are questions about the morality and ethics of performing invasive brain surgery to implant such chips. Beyond that, DARPA identified two other areas of principal concern. First, neural implants could open a new avenue for hacks that use a radically more direct form of control—give me what I want, and I'll let you have your brain back. Second, if security issues were addressed and the potential to restore or augment human senses was made available, would it become something only the affluent could afford? If rich parents can augment their kids and poor parents can't, we risk driving an even greater wedge into society. The concepts of viruses and firewalls take on a whole new level of meaning in this environment, Alvelda says, and they're not issues confined to a far-off future.

Unforeseen quandaries will arise as neural implants gain more power and wider use, too. By 2035, as more communication occurs directly from one brain to another, we might need a whole new method of verifying spoken word inferences. Meanings might change, as an official "spoken" word suggests one idea, but the concept transmitted to smaller set of people suggests another. It will get harder to know whom or what to trust. The science-fiction tropes about thought control might also come into play. Just imagine a criminal trial in which a witness is asked to testify about communications received from a defendant's neural implant. Which thoughts are private? Which can be subpoenaed? If every thought ends up being fair game, then we may end up losing our freedom to think and reflect as we choose. That is a recipe for oppression; the end of our freedom to change our opinions and the leeway to classify our own thoughts as tentative, inappropriate, self-censoring, or self-adjusting.

As Alvelda notes, it's already possible to pass information between silicon and neuron. Does every thought and interaction that moves across that threshold become fair game to investigation, depriving people of the freedom to think and reflect as they choose? If uploads, downloads, and lateral transmissions to others become recordable, analyzable, and judgeable, then the nature of thought itself has changed from a fleeting artifact of brain activity to a firm transaction. One person's internal reflection might become subject to editing, manipulation, and misappropriation by others.

Any progression toward more powerful neural implants and communications must include, at a minimum, a bright-line legal protection of thought. This will be a necessary first step to maintain the privacy of our cognitive and conscious minds, and it will justify and encourage the development of technologies that help shield our private inner selves from others.

NASTY, BRUTISH, AND SHORT*

The concepts of fairness and justice can be hard to define in the context of AI systems, especially when we try to account what's fair for an individual versus what's fair for a group. "We found there are a lot of nuances just to what the notion of 'fairness' means," says Jeannette Wing, the head of Columbia University's Data Sciences Institute. "In

---

*   Adapted from Thomas Hobbes quote "solitary, poor, nasty, brutish, and short" in *Leviathan*, 1651.

some cases, we might have distinct reasonable notions of fairness, but put them together and they're in conflict." For example, what might be fair for a group of people might not be fair for an individual in that group. When interviewing candidates for a job, a hiring manager could deliberately choose to interview unqualified female candidates simply to satisfy statistical parity. After all, granting interviews to the same percentage of male and female applicants would suggest fair treatment of females as a group. Clearly, though, by deliberately choosing to interview the unqualified female applicants, the manager would be acting unfairly toward qualified women who applied but weren't granted interviews.

The individual-group divergence matters because so many AI agents draw individual conclusions based on patterns recognized in a group of similar individuals. Netflix recommends movies based on the ways an individual's past patterns mirror those of like-minded viewers. Typically, systems can draw blunter, but more accurate, predictions from a larger group. Narrowing in on a smaller group for finer recommendations raises the likelihood of error. That's not a problem when trying to pick the evening's entertainment; it's a major problem when trying to define the terms of probation or a jail sentence for a defendant. As judges supplement their own expertise with statistical models based on recidivism, they risk losing the nuance of an individual defendant's circumstances. In theory, such algorithms eventually could help judges counterbalance their biases and perhaps close racial disparities in sentencing, but outliers will always exist and these systems, as of 2018, are not as failsafe as we expect objective machines to be.

The context needed to consistently and accurately identify these unlikeliest of cases does not yet exist in consumer-facing AI applications, either. However, as of this writing, some breakthroughs appeared imminent in enterprise applications, including at Intel Saffron. According to Gayle Sheppard, the head of the chipmaker's Saffron AI Enterprise Software unit, most fault detection in manufacturing identifies the common or probable causes of flaws and failures. Saffron's memory-based learning and reasoning solutions, a complement to machine learning/deep learning, don't just identify those likely occurrences, she says, they can drill down to find and explain the outliers—the one-off weaknesses that might cause a part to break or a system to malfunction. If successful, Intel's one-shot learning AI competency could have huge implications for higher-quality manu-facturing, but might also eventually improve AI platforms that assist individual people in their everyday lives.

Yet, the tension between what's fair and just for the individual might still clash with a broader community's interests, a friction Cynthia Dwork attempts to address with her concept of "fairness through aware-ness," which seeks to guarantee statistical parity while "treating similar individuals as similarly as possible."* It sounds simple enough as an everyday part of life; we simply *feel* something is fair or unfair, and then try to construct a valid justification for that sense. But putting those concepts into the computer code has proven far more difficult. Policies of quotas for women in politics and business in Germany, affirmative action policies for minorities in the United States, and the recruitment of lower castes into governmental jobs in India all attempt to do right by a group that is treated unfairly. And yet, each creates tricky dynamics at the individual level, particularly for those who don't belong to the group of people who directly benefit from such a program.

These sorts of attempts to balance group and individual fairness often trigger cries of injustice, such as the US Supreme Court cases argued against the affirmative action or diversity initiatives at the University of Michigan and the University of Texas at Austin. In both cases, the court generally upheld the universities' policies, although not without caveats—in part because the policies might not seem fair from an individual perspective. Often, we also conflate that sense of individual fairness with justice. Such is the case with the popular conception of John Rawls's theory of justice, in which the American philosopher argued that true justice might emerge from a balance of fairness and self-interest. Rawls put his hypothetical people behind what he called "a veil of ignorance." In this "original position," they couldn't know their own attributes nor how they might match up against the rest of the people in their society. So, as they designed principles of justice for the entire community to live by, they couldn't stack the deck in their favor. No one could know where the distribution of resources and abili-ties would leave them on the socioeconomic ladder, so everyone would tend to design a system that treats every individual fairly.

---

* Cynthia Dwork, et al, "Fairness Through Awareness," Proceedings of the 3rd Innovations in Theoretical Computer Science Conference (November 29, 2011).

Rawls makes regular appearances in AI writing and punditry, often in relation to the moral status of AI agents as they become increasingly intelligent. But his "veil of ignorance" is also worth noting as a metaphor for the role artificial intelligence might play in justice systems. Humans can never go back and place themselves behind Rawls's veil of ignorance, but in theory they might develop a set of AI systems that simulate his concepts of "original position." Such a system, if feasible, might suggest a most just weighting of interests between the accuser, the accused, and the communities in which they live. Of course, many obstacles block this theoretical pathway, not least of which is the unintentional bias that lurks in nearly every data set we compile. Yet, perhaps even an imperfect simulation might help guide a collaboration between law enforcement, victims-rights groups, watchdogs such as the American Civil Liberties Union, and community activists.

Regardless, questions of justice and fairness will only get more complicated as artificial intelligence parses out more intricate discrepancies between groups and individuals. Social backgrounds, behavioral traits, academic and job performance, responsible conduct, and past disadvantages—all of these types of data can feed into AI systems. As this mix of information and algorithm reaches levels of complexity beyond human understanding, how will we reassess who really deserves society's support? As our lives become more digital and AI analyses get more granular, will we lose the subjective context that influences our sense of justice and fairness? In the United States, for example, juries consider all sorts of contextual information about a criminal case, from prior criminal records to witness credibility on the stand. Add the constant evolution of US case law, and even our cut-and-dried notion of strict legal justice is immersed in subjectivity.

So, at the least, we might start with policies that hold the people, companies, and government entities accountable for the fair and just use of artificial intelligence that directly affects people's lives. "Generally speaking, the job of algorithmic accountability should start with the companies that develop and deploy the algorithms," Cathy O'Neil writes in her book, *Weapons of Math Destruction*. "They should accept responsibility for their influence and develop evidence that what they're doing isn't causing harm, just as chemical companies need to provide evidence that they are not destroying the rivers and watersheds around them. . . . The burden of proof rests on companies, which should be required to audit their algorithms regularly for legality, fairness, and accuracy."* "We are just arms merchants," one tech executive said in a casual conversation with us. He said it in jest, but it illustrates the tragedy of the tech sector today—the people who set out to "change the world" now slide deeper into ethical conundrums and need much stronger governance on these sorts of issues than any time or any industry before. It's incumbent upon us to make that happen.

ECONOMIC GROWTH VERSUS HUMAN GROWTH

Let's not pretend otherwise: The development of AI systems will transform work tasks and displace jobs at unprecedented speed. It also will spur demand for new skills we have yet to imagine. What we don't know is how people, governments, economies, and corporations will react during this turbulent process. Some countries might opt for an embargo on robotics, others for a tax on them. Still others will limit the amount of analysis and data that companies can keep or the depth at which AI-powered applications can permeate our daily lives. Labor organizations might revolt, with walkouts the likes of which we've seen historically among farmers in France or autoworkers in the United States. A new generation of Luddites might seek to destroy the machine or to drop off the grid and withdraw into analog safe havens. Some economists and politicians suggest a universal basic income (UBI), providing a threshold amount of income to every citizen to help support people who lose their jobs to AI and other automation. Others argue workers could use part of their income to take an ownership stake in the machines that disrupt their jobs. Truck drivers, for example, could own and profit from the autonomous rig that replaces them.

But alongside the anxious reactions and calls for safety nets against job loss, we'll also start to realize that cognitive machines are taking on many of the tedious tasks that make today's work so banal for so many. About 85 percent of workers worldwide say they felt "emotionally disconnected from their workplaces," according to a 2017 Gallup survey. What exactly are we trying to preserve? Perhaps it makes more

---

* Cathy O'Neil, *Weapons of Math Destruction* (New York: Crown, 2016).

sense to design and train workers for AI-augmented tasks that lead to increased productivity, greater stimulation, and higher purpose. Perhaps a *symbio-intelligent* relationship with AI systems affords workers more time to pursue the types of work that fulfill them and drive better results for their employers, or simply gives them more free time to be happy, sipping daiquiris on the beach.

Across all cultures and societies, people love to create things and express themselves. The rules and conventions for that self-expression might vary, but creativity and fulfillment often manifest themselves in work. Tapping that potential energy could unleash a new wave of productivity and development, lifting standards of living and innovation around the globe. Adobe Systems, the company behind Photoshop, Illustrator, and a range of software used by artists and designers the world over, has deployed AI to help remove the tedium of the creative process. From systems that automatically eliminate the "red eye" in photos to tools that can realistically swap photo backgrounds to suit the needs of an advertising campaign, Adobe's advancements allow people to spend more time on the creative parts of their jobs. Changes that once took weeks might now take a few minutes. "This enables productivity. This is the efficiency we're talking about," Dana Rao, the company's vice president of intellectual property and litigation told California lawmakers in March 2018. Creative professionals are "still using their skill," Rao said. "They're still using their intelligence, and their job just got a lot easier."

Naturally, there will be a dark side to all this. People already use AI-based creativity tools to misrepresent, defraud, or mislead. (We even say doctored photos were "photoshopped," after all.) By the spring of 2018, several digitally manipulated videos of former president Barack Obama had gone viral, all of them looking fairly realistic but none using anything he'd actually said. Some were good-natured, some not so benign, but each had manipulated his facial expressions, mouth movements, and voice to look authentic to a casual observer. Skilled digital craftspeople can put together many kinds of visual evidence, such as fake news videos, that make us believe things that never happened. They act as movie directors, but with the sole goal of changing our views, mindsets, discourse, and decisions. However, there are technologies to counter such fakes, including Digimarc digital watermarks in images or frames of video. And forgery will get harder as technologies emerge to make changes trackable. A filmmaker might put packets of movie scenes into a blockchain, which cannot be edited without leaving a trace. The use of such verification and ledger technologies—distributing confirmation and record keeping across a wide group of users—will limit the chances of manipulation and fraud.

Still, the threats of nefarious use will do nothing to slow the uptake of all kinds of AI-powered tools. Historically, and by their nature, corporations have an incentive to use machines operated by fewer workers who are highly skilled. Many nations, particularly the United States, lack the critical economic and policy incentives that would spur companies to educate and transform their workforces. From the corporate perspective, AI increases incentives to spend *less* on human labor by increasing worker productivity. Already, virtually every company of any appreciable size is thinking about how to integrate AI into its operations. Global venture capital investment in artificial intelligence and machine learning doubled to $12 billion in 2017 from $6 billion the prior year, according to KPMG.*

If historical trends are any indication, a significant portion of that investment will transform the types of skills companies demand. In 2017, an Oxford University study grabbed headlines, noting that 47 percent of US jobs are "at risk" of being automated in the next twenty years. A McKinsey report released that December said about half the world's jobs already were "technically automatable" with existing technologies. It estimated as many as 375 million workers would have to shift to new occupations by 2030, just over a decade later.† Other researchers and think tanks take a more granular view of job disruption, in some cases with different results. The Berkeley Work and Intelligent Tools and Systems (WITS) working group, in which we participate, explores how we can go about shaping the world and the workplace in an age of intelligent tools. The interdisciplinary collaboration takes a task-based view, among other things, of the technological transformation of work. A separate German study suggests smart technologies will not lead to dramatic unemployment, but will have a structural impact on task composition and employment in certain types of jobs. The impact will be negative on manufacturing, potentially positive on services,

---

* "Venture Pulse Q4 2017," KPMG Enterprise report (Jan. 16, 2018).

† "Jobs lost, jobs gained: Workforce transitions in a time of automation," McKinsey & Company (December 2017).

OLAF GROTH AND MARK NITZBERG

and will likely affect men more than women.* A few think tanks and consultancies take a more optimistic view. For example, a January 2018 report from Accenture suggests businesses, by investing in state-of-the-art human-machine collaboration, could boost revenues 38 percent and increase employment levels 10 percent by 2022.

Ultimately, the question isn't whether jobs will change and workers will be displaced—many undoubtedly will. And it won't even take a superintelligence to do it; the evolution of the narrow AI agents we already see in 2018 will automate many more workplace tasks. The question is how quickly these transformations will occur, whether we can keep up with them (especially when it comes to education and workforce training), and whether we can develop the imagination to see what sorts of new opportunities will arise with the changes. We won't run out of work, though. As Tim O'Reilly, the founder and CEO of O'Reilly Media, says in his video called "Why We'll Never Run Out of Jobs," we'll always have more problems to solve.† But adapting to the new nature of work will require imagination and preparation. The Luddites were right that industrialization threatened their lives and well-being, but they didn't have the imagination to see beyond the initial disruption. Most companies see the work before them and what a thinking machine can do better and more cost-effectively, but they don't look ahead at the skills the workplace will need tomorrow—things like man-machine teaming manager, data detective, chief trust officer, or the eighteen other "jobs of the future" that Cognizant laid out in a 2018 report.‡

Yet, imagination only goes so far, even when equipped with stockpiles of resources to facilitate it. If US companies needed nothing more than cash to prepare themselves and the American workforce for this future, they would've already been doing it with the $1.8 trillion in cash reserves they had on hand at the start of 2017.§ Companies have valid strategic rationales for retaining this much liquidity, which

---

* Researchers studying in this vein include: Carl Frey and Michael Osborne (Oxford University); Wolfgang Dauth, Sebastian Findeisen, Jens Südekum, and Nicole Wössner (Institute for Employment IAB); and David Autor (MIT).

† https://www.oreilly.com/ideas/why-well-never-run-out-of-jobs-ai-2016

‡ *21 Jobs of the Future*, Cognizant, Nov. 28, 2017.

§ *Moody's: US corporate cash pile grows to $1.84 trillion, led by tech sector*, Moody's Investors Service, July 19, 2017.

allows them to respond quickly to disruption and fuel research and development for new products and services. However, investments in innovative concepts for the future of working, earning, and learning will require a long-term focus on the needs of global and national economies. Driven by short-term shareholder interests, few companies have an effective incentive to imagine an undefined future.

Policy makers could choose a wiser course aimed at achieving both greater productivity and competitiveness for corporations while getting our workforce ready for the Fourth Industrial Revolution. For starters, they could create incentives and encourage public-private partnerships that spur corporate investment in the development of and training for defensible jobs of the future—in fields such as clean energy, technical design, and 3-D manufacturing. Governments could consider similar incentives for investment in civil and business infrastructure, including innovative transportation solutions and the revival of older manufacturing hubs for a new economy. And they could apply the same incentive logic to investments in affordable housing, so San Francisco, Shanghai, Berlin, Mumbai, and the other global economic hotspots are welcoming to more aspiring workers.

Unfortunately, in their current forms, few national strategies are doing anything to get workers there. So, we also need to educate people for those jobs, many of which will require skills or combinations of skills unheard of in today's workplace. Public-private partnerships could define the outlines of future job categories, build hybrid online/offline training models with project-based learning, and offer credit-based upskilling programs with nano-courses and education certificates. They might create integrated corporate apprenticeship programs of the sort that German companies, such as BMW and Volkswagen, have developed at home and have brought to the United States. Workers with lower skill backgrounds could earn extra tax credits for participating in these programs and taking the courageous step to upgrade their skills, perhaps even receiving a universal basic income to support them as they go.

With a combined corporate-and-labor "relief and reskill" program, governments could transform the existing mindset of competition between humans and technologies and move toward greater, integrated productivity through future-resilient jobs—and thus establish themselves as trailblazers for a high-tech future that enables human potential. But even within the structures of today's workplace, cognitive technologies might help spur greater productivity for companies and greater rewards for workers by providing more insight into the murky recesses of our motivations and intentions. They might even put our own subconscious to work on our behalf. For example, One2Tribe, a Polish start-up led by Wojciech "Wojtek" Ozimek, helps clients motivate employees with an AI platform that analyzes personalities and then provides rewards to encourage more sales or better call resolutions. The firm employs a mix of psychology and computer science expertise, but one of the biggest insights came by simple trial and error. Unless workers can opt in or out, they object to a system that nudges behavior in such a personal way, Ozimek says. So, One2Tribe requires that its clients only use the system on a voluntary basis. The rewards typically get about 60 percent of eligible employees to participate, he says.

The platform works much like a flow model in video games, carefully balancing the challenge with the reward, Ozimek explains. But it goes a step further, with its psychology experts testing everything from real-time responses to actual brain function, so they can better understand the challenge-reward relationship, identify the most effective approach for each individual, and then update it on the fly. The timing between challenge and reward is especially crucial, he says. One worker on a task might produce better results with a larger weekly goal, while another might produce more when getting smaller rewards on a daily basis. The system typically distributes a sort of virtual currency employees can exchange for other items. "We create AI to balance goals with demands," Ozimek says. "We take into consideration the skills of the person, their personality traits, and then we try to create a motivational scenario."

Of course, the subtlety and depth of One2Tribe's influence on worker behavior naturally spark concerns about manipulation. The platform worked poorly before Ozimek and his colleagues realized it had to be a voluntary option for workers. But even on a voluntary basis, safeguards are needed to ensure that companies don't deploy similar AI-powered systems without checks and balances. Our future need not include extrinsic incentives and the gamification of rewards that play us like organs—or worse, treat us like machines that produce without fulfillment and purpose. Individuals, society, and the planet need us to think about what's right from the inside out, not just the outside in.

Yet, we also need acknowledge that rank-and-file workers will play a key role in creating safeguards, as well. A company that tries to engineer its employees' mindsets will not create a desirable workplace that draws the best talent. It's not for nothing that Glassdoor has become a go-to for workers to rate and review workplaces, or that businesses tout their rankings on *Fortune*'s 100 Best Companies to Work For. But we undoubtedly will experience a lot of push and pull on worker-related AI systems as we calibrate the different types of employee stimulation. Companies will risk crossing the line at certain points, whether purposely or inadvertently.

Responsible organizations will want to track performance and identify areas of potential abuse. Progressive businesses will want to make that as transparent as possible. This might begin with a deliberate cooperation between labor and employer, collectively creating the rules for systems that influence worker behavior. This might resemble the existing joint efforts of labor and business in Germany, where both sides are working together to guide the deployment of robotics and worker training programs. It might eventually include certification of worker-related AI platforms by a professional group, such as IEEE, or internal and industry-specific labor review boards that can audit such systems. Regardless, companies and government need to take into account morality and professional codes of conduct to mitigate bottom-line myopathy, especially in economies driven by short-term quarterly results and the stock option packages tied to them. Maybe then, as we prepare for this unfolding cognitive revolution, we can assess whether we should include stakeholder ethics in calculations of performance-based pay.

HUMAN GROWTH IN HEALTH CARE

For many people, the subtle nudging of our motivation and mindset might seem just a little too intimate, even when only offered in limited settings and on an opt-in basis. The use of AI in our physical and mental health care might feel even more invasive, one of the reasons human doctors have remained central in most of the health-related systems to date. The power of AI in health care lies in the fact that cognitive machines can process myriad data streams and recognize complex patterns much faster and with greater accuracy than human brains can. Image recognition systems now surpass expert human performance on many radiological tests. IBM Watson's ability to process reams of cancer-research literature in a week, and then learn virtually all the potential therapeutic techniques the next week, lies far beyond any collective human capability. Putting those sorts of cognitive machines into symbiotic partnership with trained doctors and other AI-powered systems—cognitive networks that understand the vast multitude of pharmaceutical treatments and their side effects, for example—could advance medical care and human well-being to unprecedented levels.

Yet, all of this still maintains the central role of human doctors, nurses, and medical lab professionals, who combine scientific meaning, socioemotional structures, and the mindsets of patients into a harmonious and effective delivery of care. "We'll still need radiologists to explain things, to take those findings and explain them," says Clay Johnston, dean of the Dell Medical School at the University of Texas at Austin. "But the vast majority of hours spent by radiologists today will be taken over by machines." Similarly, smart machines might eventually recognize emotional states by way of facial or voice recognition algorithms, notes Jonathan Gratch, a computer science professor at USC, but accurately interpreting human goals and agendas is harder than many researchers realize. Most current approaches assume that recognizing surface-level expressions of emotion, such as vocal tone or facial expression, will be sufficient for understanding a subject's mental state. However, Gratch says, systems will need to interpret those surface cues in context, because people often mask or misrepresent their expressions. If a poker player smiles, it doesn't say much about his or her mental state. If that smile appears after a new card arrives, it might suggest a lot more about how he or she is thinking—or it all might be part of an elaborate bluff. It's a messy and complex problem.

Nor can machines truly experience empathy, perhaps one of the most critical dynamics for a successful doctor-patient relationship. An AI might simulate empathy, and sometimes that's enough to stimulate more frank and open responses from patients, particularly in cases of mental health. But ultimately, doctor-patient relationships rely on a reciprocal trust that patients will honestly explain their ailments and that doctors will maintain the highest practical threshold of care in diagnosis and treatment. That mutual trust embodies the shared experience of millions of years of human evolution. A good doctor knows just how keenly pain, ignorance, and embarrassment might warp a patient's recitation of symptoms. Most patients rely on the knowledge that physicians share those human foibles, understand them, and know how to dig beyond them to find the core problem at hand. That common human experience allows a deeper person-to-person understanding that a narrow AI—with its findings based on patterns across groups—can't share.

Yet, by identifying far more complex or subtle patterns across groups, AI systems can find problems human doctors, radiologists, and other health care professionals can't spot. Integrating that penetrating objective analysis with human empathy can generate far deeper insights into our health and well-being. But that combination will not happen until AI systems are accepted and put into use by medical professionals. That's no easy task. First, there are so many human variables that play into our health-related decisions. An adult might decide to just treat their moderate fever at home, because we know how to manage it, but a first-time parent might reasonably run to the ER when their child's temperature rises. "There's an infinite number of ways that a finding can be misinterpreted by an individual," Johnston says. "I think it will take a long time, or a longer time, for computers to grapple with all the nuances of the human interactions with the facts."

That might also include the biases and misinterpretations that doctors bring, being humans themselves. Placed into the workflow of a clinical setting, an advanced AI system might notice anomalies in how a physician approaches different patients—anything from a different conception of patients' pain tolerances to a different expectation for their adherence to a treatment plan. If such a platform was available when Ann and I (Olaf) met with the first cancer specialist in Berlin, it might have recognized that his advice to terminate her pregnancy was based on his experience with losing his own wife to breast cancer. That background might have convinced him of the need to act decisively and quickly, rather than considering a riskier alternative approach.

Such a platform will take years of additional innovation, but it could take years before doctors fully integrate today's emerging AI technologies into their daily workflows. It might seem simple enough, but in complex and heavily regulated health care environ-ments, doctors have shown extreme reticence to work provably better technologies into common use. Johnston still recalls when oxygen saturation devices came out—simple devices that fit on the end of a finger and measure oxygen levels in the blood. At the time, many physicians said it wasn't enough. They said doctors needed full blood-gas workups, which you could only do once every day or two, to properly track gases. Yet, the oxygen monitors have had a dramatic impact on saving patients in the years since they were introduced.

The reticence gets even more mundane, Johnston says. Something as simple as email remains radically underutilized for patient-doctor com-munications. That will only begin to change when nonuse hits physicians in the wallet, or when an advanced technology fits into their workflows, rather than requiring that they adapt to it. For all the work on health-care AI systems today, precious little work is being done to make sure these systems work for the physicians who would use them. Johnston and his colleagues at the Dell Medical School have piloted a language-processing system that allows doctors to talk with patients, and then not only convert the speech to text but properly populate the information into standard insurance or clinic forms.

Other companies have taken similar approaches to making the machine fit the human, including an Israeli start-up called Aidoc (pronounced "aid doc," despite the unintentional play on words, says CEO Elad Walach). The company has jumped into the rapidly transforming radiological imaging space, but two factors set it apart from competitors, Walach says. First, its approach is com-prehensive, identifying a wide range of abnormalities rather than analyzing scans for a single or small set of ailments. Those focused approaches work well, but they don't allow for Aidoc's second advantage—that its results are easily integrated into a radiologist's or physician's day-to-day activities. Rather than different tools to identify different ailments, this one system red flags a wide range of problems, so it's easier to integrate into the day-to-day workflow, Walach explains. A decade from now, the health care environment will be far different, and perhaps then doctors in Israel, Europe, and North America will readily adapt to different AI and other advanced technologies. "But now, to penetrate and get traction in the market," he says, "we have to respect the place of physician and give him added value in his work."

THE CYBER BLUES

We have seen unprecedented waves of cybercrime and online terrorism in recent years. From the National Security Agency's Stuxnet virus, which brought down an Iranian nuclear plant by getting its uranium centrifuges to overheat, to the theft of millions of personal records at Target and Equifax, these incidents have become an almost commonplace occurrence in our lives. As we brace ourselves for a big, infrastructure-disabling breach of our digital economy, we seek security in an opaque and often shady arms race with illicit individuals and organizations around the world.

To be sure, the concept of AI safety involves several related pursuits.* But while the cybersecurity experts who work on the front lines of this battle believe we can limit the damage and try to reduce the ripple effects, hackers have always had the upper hand and always will. Networks have so many potential access points. The hacker only needs to find one weakness one time; a company or individual has to protect all those points of entry every second of the day. The development of AI-powered cybersecurity applications might provide faster reactions and better coverage, but it's by no means comprehensively effective, experts say. And, meanwhile, nefarious actors will be developing and enhancing their own AI-powered hacks.

"We realized very early on that the industry as a whole was thinking about preventing attacks from happening," says Yossi Naar, cofounder of Cybereason, an Israeli cybersecurity start-up. "We realized experienced attackers can always get in, and the trick is finding them inside the environment." Cybereason patrols the "endpoint" of the network, the outer edges of human-computer interfaces where attacks enter. By watching those edges and using machine learning to analyze fine usage patterns down to an individual and device level,

---

* Safety in the AI context can refer to 3 different things: 1) Safety-purpose AI, or technology built *to solve a safety problem*, such as crash-avoidance for cars; 2) Cybersecure AI, or technology that *resists adversarial compromise, unauthorized alteration, or control*; and 3) *safe AI* that is built from the ground up to avoid flaws and undesirable outcomes.

Cybereason can identify anomalies that don't fit within a chain of events or typical patterns of behavior and more quickly react to them. Watching the far edges of the network provides the most comprehensive data set on usage, but the massive wave of information this produced made such monitoring impractical before cheaper, more-powerful computing emerged to help machine learning systems process the torrent.

At the Silicon Valley start-up DataVisor, cofounder and CTO Fang Yu doesn't really expect a bigger or better wall to keep hackers out. But by creating an unsupervised learning system that analyzes the millions of legitimate transactions conducted across a client's network—and then using that knowledge to identify oddities, including previously unseen types of attacks—the company's technology can stop more of the bad guys, Yu explains. Other systems require training data or labels to teach the system what to watch for, and then it looks only for just those things. Hackers will launch "massive attacks all of a sudden because they can hit many accounts at once," Yu explains. "An unsupervised algorithm is able to detect the new pattern forming and say this group of accounts is very similar in terms of behavior and it's very different from normal users."

Hackers can mimic one account or a small group of them, but the large-scale fraud that can take down a company creates its own pattern. Identifying those without prior training allows DataVisor to react more quickly and eliminate the false positives that can run up costs and decrease the effectiveness of cybersecurity measures. In one case study, the firm's system helped increase detection of account takeovers at one of the world's largest online payments platforms by 45 percent. False positives dropped to 0.7 percent.

Still, the sad truth of the matter remains that hackers have so many options to get in, both digital and analog. Simple human deviancy and disenchantment will work just fine. Given the asymmetry between potential attack avenues and the difficulty of defending them, a much larger crash is almost inevitable, says Ivan Novikov, CEO of Wallarm, a San Francisco-based cybersecurity firm. Companies typically take two approaches to detecting attacks. Traditionally, they would hire security analysts to analyze samples of malware or malicious traffic and create "signatures" based on those examples. The signature might include a unique section of code or combination of elements that identify it as toxic. When new attacks arrive, the process begins anew.

Wallarm uses neural networks to do something similar, but it creates statistical profiles instead of signatures, and it can develop and deploy them in real time, Novikov says. That provides greater security, but given the unending and largely unwinnable battle against malicious hackers, he also has resigned himself to the fact that we'll see far worse attacks in the future. "I can't predict when and how," he says, "but I expect to see a global Internet shutdown in the next five to ten years. People already tried last year with botnets. So yeah, we'll see a global Internet shutdown with a significant amount of Internet service being unavailable to a significant number of users."

Extend that sort disruption to the interconnected infrastructure, and serious crises could emerge. A blacked-out power grid doesn't come back up with the flip of a switch. A widespread outage that stretched into a week or more threatens hospital care, food availability, security, heat and air conditioning, and the maintenance of so many other critical systems. Air traffic grinds to a complete halt, and road traffic descends into gridlock. But while nobody is served by minimizing the dangers, we shouldn't forget the saving graces available to us. DARPA's initial iteration of the Internet established a naturally resilient communication infrastructure, one that could survive an attack in one part and automatically reroute traffic to another. Every computer connected to it can function as a node—if a million computers are infected, millions of others could pick up the slack.

Researchers also have started developing defensive AIs that could neutralize malware as it enters our most important network nodes. That said, the threat to the Internet and infrastructure should prompt us, as individuals, to create redundancies on both the digital and institutional facets of lives, spreading our assets and critical life functions across wider areas of personal networks. Rather than staring into the abyss, we can prepare and make our own, personal infrastructures more resilient. Many people already do this, if for other reasons, for their own homes, adding solar panels and rainwater capture systems. We might need to build the same backstops for our digital lives, setting kill switches that trigger when the "Trojans" appear, and then rebuilding our bridges after we've run them off.

For example, it might make sense for every smart home to have an "analog island mode" that switches off all its digital connections and seamlessly transitions the home to a safe operating mode without interrupting critical functions. This could safeguard the family's life, protecting critical hardware such as sleep apnea machines, baby monitors, refrigerators, and home alarms. This type of safety switch might also protect the broader power grid, helping balance electricity load in the event of an attack and avoiding costly blackouts. Communication and cooperation between grid operators and the many homes, commercial spaces, and factories that rely on a secure electricity supply might help contain outages or more quickly recover from those that occur, whether sparked by overgrown vegetation or a nefarious attack.

THE POWER OF CREATING A BETTER TOMORROW

Israeli entrepreneur Yaron Segal spent the better part of the past decade searching for a better way to help his son, who suffers from *familial dysautonomia*—a debilitating syndrome that affects the types of nerve cells that control involuntary actions, such as digesting, breathing, and producing tears. As a father and a scientist, Segal felt compelled to discover the fundamental aspects of his son's malady. That search took him to other brain injuries and neurological disorders that share some of the same basic properties as *dysautonomia*. And, having identified those, he began to figure out ways to decrease the effects of those problems. BrainQ, the Israel-based start-up, was born.

Hundreds of millions of people around the world suffer from neural disorders, and the cost of care keeps escalating, typically in the form of physical therapy and living assistance costs, says CEO Yotam Drechsler. The problem with many of the current treatments, Drechsler says, is the fact that they treat the symptom, addressing a partial paralysis of an arm or leg instead of addressing the injury or limitation that stems from the brain or spine itself. Since the initial team consisted of both traditional and data scientists, they looked at the problem in a fresh way, combining cutting-edge machine learning methods and tools with scientific theory, particularly Hebian theory, which claims: "Cells that fire together, wire together."

The company put the Hebian theory to work in a novel way, hoping to re-create neuroconnectivity in damaged or limited portions of the brain. "Why don't we do physical therapy directly to the brain?"

Drechsler says. "We believe there are specific patterns in the brain that are associated with every movement. And once we identify them, our goal is to imitate them with a non-invasive device based on a low-electromagnetic field, and hopefully achieve this neuroplasticity. We learn to imitate the 'firing' patterns with advanced AI tools in order to facilitate the 'wiring.'" While neuroscientists could understand how certain networks controlled certain bodily functions, measuring precise firing rates had been all but impossible with traditional technologies. BrainQ developed a machine learning algorithm that, it says, provides clarity on the disrupted frequencies. The company now claims to have one of the world's largest sets of electroencephalography (EEG) motor task data for Brain-Computer Interface (BCI) applications. "The firing rate is not the new story," he explains. "The problem is people were trying to learn it for years and did not find much because the signal-to-noise ratio is so low. The data is extremely noisy, so it requires sophisticated tools such as BrainQ's to interpret it."

With those signals isolated, the company uses a coil, which looks like a salon's hair dryer, to provide a low-intensity signal at the right frequency, reigniting neural firing in hopes of re-creating neural wiring. The company has a video that shows a partially paralyzed rat that can't move its back end and drags itself around by its front legs. After a month of treatment, tests showed a partial connection of the damaged tissue and the mouse is moving its back legs, albeit slowly. By day fifty-eight, the rat is climbing up boxes and its back legs appear fully functional again. The treatment uses low-intensity energy, so it poses no known physical risk, and it can treat a variety of brain maladies. BrainQ has run some limited human trials in Israel, and those tests have shown promising results. And, as of this writing, the company was beginning to explore opportunities in the United States, too. Could this help treat the 250,000 to 500,000 people around the world who suffer spinal cord injuries each year?[*]

Developers, scientists, and researchers are using artificial intelligence to tackle some of humanity's and the planet's most confounding challenges. Some of them, like BrainQ, are using these powerful AI systems to engineer treatments heretofore impossible, discovering

---

[*]   *Spinal cord injury fact sheet*, World Health Organization, Nov. 19, 2013.

new avenues to solve known challenges. Others, like Yasuo Kuniyoshi, are seeking a deeper understanding of intelligence and learning itself. Kuniyoshi is the director of the Intelligent Systems and Informatics Laboratory at the University of Tokyo, where he and his colleagues experiment with complex neural networks and robotics in hopes of developing more capable AI agents—and they're learning a little more about human development, as well.

Kuniyoshi has developed a robot fetus, a complete robot and computer simulation of a developing human body and nervous system. The extremely precise model even floats in a liquid-filled "womb," in which it wiggles and moves spontaneously. Through sensors and its neural network, the robotic fetus begins to learn about itself and its environment. So, if its limbs touch each other, he explains, the robot recognizes and learns about that physical relationship. "That, in return, changes the output," he says. "As the baby learns it changes its behavior and changes its input, because the movement changes. If this continues on, we think we can call it a spontaneous development."

It's still in the early stages of development and understanding. The current systems are still relatively small, with a few million neurons. But no one knows what level of complexity would trigger, say, self-reflection or similar human abilities. That would require massive amounts of compute power and further research, but Kuniyoshi is convinced the relationship between the neural and physical—the cognitive computer and the robot—is indispensable to understanding how and when those attributes develop. "I firmly believe that for the future of AI, human likeness is really important," he says. "A human-like mind is really grounded in a humanlike body and a humanlike environment, and the initial trajectory is very important."

Shreya Nallapati might not choose to follow Kuniyoshi into the depths of advanced AI research, but she is clearly on to something else that is important to our well-being. Before we spoke with her, she'd settled into her senior year of high school in Colorado, thinking about where to go to college and pursue her interests in computer science and cybersecurity. Then, on February 14, 2018, nineteen-year-old Nikolas Cruz killed seventeen of his classmates at Marjory Stoneman Douglas High School in Parkland, Florida. Like previous mass shootings, this one reignited the debate about gun control, but it also sparked a new and larger movement—one that was launched by the victims' classmates but quickly spread to high-school students and young adults across the country. Inspired by Emma González, one of faces of the emerging #NeverAgain movement, Nallapati founded #NeverAgainTech, an initiative that applies machine learning, natural language processing, and other AI technologies to identify the root causes of mass shootings and, hopefully, prevent them in the future.

Supported by organizations such as Amy Poehler's Smart Girls, Forbes, and technological contributions from some Silicon Valley companies, including Splunk, Nallapati and her collaborators began to identify various factors that contribute to mass shootings. They looked at a variety of data that fit into five primary categories: mental illness in the shooter; their socioeconomic status and social/familial background; their motive; the firearm used; and any state policies on firearms. The goal of the all-girl team, Nallapati says, is to come up with an ironclad, evidence-backed argument for national or state policies that can overcome entrenched political views and bring together industry professionals, students, and policy makers. "At the end of the day," she says, "we realize that although we cannot actively prevent the next shooting, we can do our best to identify potential hot spots for such activity and shed some light on a very complicated and emotional issue."

The mere mention of artificial intelligence can trigger anxiety about all the ways things could go wrong and about how these subtly powerful technologies could be used to harm humans. Yet, these AI systems are still tools, ones that can make the world a far better place if we proceed with care. We will need to strike a balance of power between individual agency, governments, and the world's Digital Barons. We will need to shore up the existing institutions that provide the safe and secure flow of electricity, water, and data. And we will need to preserve trust and human-centric values.

Segal, Drechsler, Kuniyoshi, Nallapati and thousands of like-minded people around the world are tackling some of these challenges already. Whether each one ultimately succeeds in his or her quest might not matter to the world at large, but the care and passion embodied in their visions for AI will make all the difference for the future of humanity.

7

Life and love in 2035

CONNOR (THE EX-BOYFRIEND)

Connor reset the plow and stopped to check the harnesses on Whoop and Holler. The mules had been at it all day, and now they lined up toward the dusk gathering beyond the far swath of birch and maple. Though they were tired, their ears pricked up as he came around, and Connor felt his usual compulsion to finish the job. So, he wiped his brow on the back of his leather glove, took a moment to admire the consistency of his lines, and then stepped back behind the plow.

"Why can't you just let it go?"

The voice flittered out from the shadows of his mind, and the queasy mix of anger and longing settled into his stomach. Ava's voice—he heard it more often lately. At first, the sounds of his ex-girlfriend's voice infuriated him, spinning him into bilious frenzies of work. The community didn't know what set him off, but they knew what happened whenever it did. After more than a year of doing as little as possible under community rules, Connor suddenly started checking off the entire list of daily errands by midafternoon, if not earlier. By now, the rest of the Siblings knew he wouldn't stop until every last bit of each job was done, and done well. Whenever he set off on a fever, they just eased back and let it play itself out.

It didn't happen as much lately. Connor still worked as hard as anyone there, but never with the same sort of madness. Even now, as he exhorted Whoop and Holler up the next row, he spit his usual mantra at no one in particular—"Why the hell would you agree to do something and then not do it right?"—but he sighed as soon as he said it. He was back in his own head, and his head was back in the apartment he'd shared with Ava. He'd mutter the same thing to himself, just loud enough to make sure she knew something was annoying him: *Why don't you just put your shoes in the closet where they belong?* It was never a big thing, until it was, and by then it was too late anyway.

"Why can't you just let it go?" Ava said when Connor brought it up on the way to Leo's wake. They had very different reactions to their friend's suicide. Ava saw it as a calling to express herself more honestly; Connor saw it as the worst outcome of the pervasive use of personal AI assistants. He blamed the PAL for aggravating Leo's depression about his sexual identity. "Leo didn't kill himself," he thought, "the PAL did." And, as they rode over to the celebration of Leo's life, he couldn't help himself. Bitter, he asked again: "Do you really think that goddamned thing is good for you?"

Thinking about it now, behind the plow, the disgust returned. The first time Ava's PAL suggested that her biometric and behavioral evidence hinted toward bisexuality, Connor laughed it off. As she started to muse more about it, he saw it as pure criticism. By the time the driver dropped them off at Leo's wake, her two passengers were yelling at each other. "Why can't you just let it go?" she asked again. It was on that sidewalk, on that horrible night, when he decided to leave.

He ended up here, at the end of a quarter mile, ruler-straight row of plowed field in Alberta, Canada. The Community of People had emerged from a crossbreeding of anti-AI humanist intellectualism and the Amish adherence to austerity. Connor scoffed at the commune in its early days; it went too far to cut off all connections to the outside world, even electricity and phones lest some AI-influenced technology leak its way in. But he was always intrigued by the idea. By the time it had suffused through his mind enough to suggest it to Ava, the community's modestly relaxed restrictions—it now allowed certain devices, but still restricted digital communications—made the switch more palatable, at least in Connor's mind.

He left the day after the wake, after Ava had gone to work. By the time he landed in Canada, he'd reset his PAL to Bouncer Bot mode, instructing it to alert him only when his ailing mother or autistic but self-reliant brother tried to contact him. For health reasons, he kept the medical and bio-physical outputs on, measuring the tiniest nuances of his gestures and app usage, even though he knew it only could upload on the few occasions he got to town—and even though he knew it would send more information than he really wanted to share. And then he set out to convince himself that this was the only authentic, personal, and "real" way to live.

He rarely left the community during the first, angry year. He'd bunked for a few months with a guy who helped him acclimate while he built his own cabin. The next few months, he hosted a new arrival, with whom he initially clashed but eventually came to trust. It was this guy who convinced him to reengage with PAL's therapy-bot functions, and it was no coincidence that Connor's tolerance of his temporary roommate improved afterward. By the time Connor had his cabin to himself per-manently, he had settled into and had come to appreciate the routine of communal life. He found himself apologizing for fits of anger, but everyone seemed to shrug them off and laugh about how much work Connor would do when he got so agitated.

That wasn't going to happen tonight, though. The gloaming had settled across the plowed fields and the chorus of insects were humming their song of survival. The rest of it, maybe half a morning's work, would have to wait until tomorrow. He would let it go this time, he said to himself with a chuckle that felt almost refreshing. He left the plow, led Whoop and Holler back to their feed and water, and, checking his watch, real-ized he had enough time to take a shower and still catch the end of the community dinner.

He changed his mind about halfway through a cool shower, when he suddenly felt an urge to go into town the next morning. He'd missed connecting with his brother the prior week, and he figured a little rest would do him well during the hill climbs on the twenty-one-mile bike trek. So, rather than join the card games and the conversation, he noted his plowing progress on the errand board and went back to settle in for the night.

As usual, his PAL kicked into action during his bike ride, about two miles outside of town. That "Welcome back, Connor, been awhile!" always jarred him out of the meditative state of rhythmic pedaling, but never enough to make him want to change it to something else. There was still something he craved about the virtual connection to the outside world, and as he cruised down the last half-mile stretch to the outskirts of town, his PAL would give him an update on the news of the world and the latest mundane details about his mother or brother. Connor stopped at a few stores to fill his pack with provisions and a few special requests from other Siblings, and then headed for the pub for his favorite indulgence—a buttered steak, a beer, and a piece of cheesecake.

"Connor!" the waiter yelled when he walked in, and Connor strode over and gave him a bear hug. As they exchanged pleasantries, Connor sensed someone staring at him from across the room. The guy just didn't quite fit, what with his expensive black leather pants a denim jacket that looked just bit too clean and pressed. *Damn*, Connor thought as Tony took the usual order, *that guy seems familiar.* Now he was walking right toward them.

"Tony," Connor said, the epiphany finally upon him, "this sonofabitch right here is Vladimir, one of my closest friends from undergrad. Holy cow, Vlad, what the hell are you even doing here?" Last Connor saw him, Vlad was working in IT at a San Francisco nonprofit. They both laughed and jumped into a deep, hearty embrace.

"Your brother said I'd find you up here," Vladimir said. "You'd all but dropped off the face of the earth a couple years ago."

"Well, there's a heluva story about that," Connor said, collecting his beer and nodding back to the table where Vladimir's double-oaked bourbon awaited.

The steaks gone, a third beer in hand, and almost two hours passed, Connor waffled between bitterness and despair as he told Vladimir about how things ended with Ava. Somewhere his inhibitions warned him that he no longer drinks as much as he once did and might want to cool it, but he drifted further into the window of darkness. He told Vladimir about how Ava's PAL suggested she slow things down with Connor, and how they decided to push through—but it wasn't the same after that.

It got worse and worse, and, after the wake, he'd had enough. He just couldn't let it go.

His mind only got darker when Vladimir told him about Ava, who was now seeing Emily, a moderately famous conservative pundit. "A right winger, eh? Well that just figures. Maybe we need an AI to split them up, too, for everyone's good." It felt good to say it, but Connor still regretted it as soon as it left his mouth. He worried even more when Vladimir flashed a thin smile and quickly changed the subject.

"Look, I gotta get going, man," Connor said. "Believe it or not, I have to go get the mules and finish plowing a field."

Vladimir laughed, and Connor couldn't help but join him, both getting a kick at just how absurdly life had changed for both of them over the past fifteen years. They embraced again, and Connor waved goodbye to Tony as Vladimir paid the bill. He hopped on his bike, turned northwest, and started back toward his new world.

"Excuse me, Connor," his PAL said, "but I have a few quick things to note. May I proceed?"

Connor assented, and the PAL went on: "First, your vitals and biometric readings look great. The cleaner living and eating is improving your health. Second, you did not call your brother. I could try to reach him now, but it appears he's in therapy. We'll have to wait ninety minutes or try again another time."

"Shit," Connor said. "Send him a note and let him know I'm good, and that I'll try to reach him in a few days unless he needs me before then. Oh, and tell him I saw Vladimir. They always got along pretty well."

"Consider it done," the bot replied. "Now, one other thing. It's been three years since Leo died, and Ava is throwing a celebration to commemorate it."

Connor caught himself just before the front tire slipped out from under him. He'd noticed his PAL probing a bit more during their therapy conversations. It asked more about his past life, confronted him about looking at old pictures, and wondered why Connor might be thinking more about Ava—much like he had done the day before while plowing.

He rode in silence for a couple minutes, marinating in the news of the invitation.

"She sent you an invitation," the PAL said. "I think you should hear it. Would you like me to proceed?"

Connor rose out of the bicycle's saddle and pushed up the steepening incline ahead of him.

JULES AND GABRIEL (THE MOTHER AND FATHER)

She caught a faint whiff of her daughter's perfume—it must've rubbed off on her jacket when they hugged at lunch—and Jules chuckled again. It never failed to amuse her, the idea of Ava's PAL recommending a rose scent she never would've worn otherwise, and has worn exclusively ever since. She ribbed Ava about it for a while, but her daughter was just as humored by the idea as she was, and they'd both end up laughing about it. "It suits her," Jules said to a golden retriever tied outside. She leaned over, gave the dog a quick scratch behind the ears, and then turned to walk downtown.

She checked her phone and saw a couple messages from her husband, Gabriel, and her other daughter, Willow. As seamless as communication technologies had become, she still preferred a good old-fashioned text message, and she still turned off her phone every time she sat down to talk with someone. She quickly flipped by Gabriel's message; he just wanted to make sure she knew where they were meeting with their retirement planner. She stopped to read Willow's message. Her younger daughter was struggling to figure out what to do next with her career, and despite the quick burst of terse messages—in fact, because of them—Jules knew she really needed to come by for one of their regular "whine and cheese" nights. She asked her old Siri assistant to suggest a couple nights the next week, and then asked Willow if she should invite Ava, too. Jules figured she'd have plenty to tell both of them.

They're not going to like this idea, she thought, and she felt her anxiety rise when she saw Gabriel waiting outside the office.

"How was lunch?" he asked, leaning in to give her a quick kiss.

"Great! Ava says hi. And she told me to tell you those shoes are a crime against humanity."

Gabriel erupted in laughter, then cut himself off with a look of faux outrage. He'd bought the shoes the week prior, out shopping with Ava, who tried in vain to dissuade him from blue suede shoes. He couldn't help but revel in the Elvis lore of it all, and started singing "Love Me Tender" right there on the sidewalk, curled lip and all. He always knew how to raise Jules's spirits. He'd carried her through her breast cancer diagnosis, the terrible experience with the first oncologist, and then through Ava's improbable birth just months later. And now, worried about their long-term retirement and care plans, he eased her anxiety again. They both were nervous, but at least they could go in with a sense of optimism.

Their planner pulled up the options they discussed on a panel embedded in the wall. He knew better than to use the hologram projections, which disturbed Jules the first time they met. Some of his clients liked the 3-D renderings and the ability to walk around and experience the care settings or interact with the simulated robotic caregivers. Jules was not one of those clients. So, he brought up the video of the HomeCare 3T bot, a common option for most of the septuagenarians these days. Gabriel loved the idea as soon as Willow suggested it—stay in their home longer, have a greater sense of security, and a little bit of companionship. "Sort of like having a grandchild in the house!" he'd joked after Thanksgiving dinner, laughing at the side-eyed glances that prompted from both daughters.

Jules wouldn't go for it, though, and all of them knew it. She craved human companionship and, despite the surprisingly natural interactions she'd had with the AIs and robots Willow's entrepreneurial friends developed over the years, they could never be human enough for her to connect with. Now sixty-five, after a lifetime of doctors, checkups, and every iteration of health care technology you could imagine, she wanted a flesh-and-blood human to talk to and rely on. Besides, the HomeCare 3T never did live up to the hype. The trade wars over the previous decade limited investment in US robotics, and the care robots in China and especially Japan had far exceeded the capabilities of the HomeCare line.

The adviser saw the sour look on Jules's face and called up the next video. The new scene opened on a lake in the mountains, a dirt road running beside it, and tall, vividly green bamboo swaying gently in the breeze. A gray-haired couple walked slowly, hand-in-hand along the road, smiling as they chatted. The video cut to a close-up, she leaned over and rested her head on his shoulder, and they both gazed contentedly into the distance. "It's perfect! We'll take it!" Gabriel said, bursting out in laughter.

"Yeah, a lot of this video is pretty ridiculous," the adviser laughed. "The hologram gives you a better sense of it, but let me get the tour of the facilities here on the screen."

This is what it had come to: They couldn't afford the soaring cost of human care in the United States. That became an unaffordable luxury for most people after insurance companies started insisting on less expensive, rudimentary robotic companions and wouldn't reimburse anything else. Most retirees had very few options after that. They could try to take care of themselves as long as possible, hoping to reach a point where they could afford a briefer stay with human care at the very end of their lives. They could settle for the mediocre HomeCare 3T and remain in their homes indefinitely. Or they could join the rising wave of "retirement tourism," going to China, Germany, or the handful of other countries that managed to contain costs while remaining human-centric. These health-care havens had invested early in higher-quality robots that augmented the skills of their highly trained immigrant elder care workers.

The adviser noted some of the amenities of the Chinese facility on the screen. He made a point to tell Jules and Gabriel that the staff didn't look Chinese because, in fact, they weren't. Despite its massive population, China's firms couldn't recruit a big enough domestic staff. So, the industry recruited and trained thousands of caretakers from the Belt and Road countries, much like German facilities did with the influx of Syrians and Iraqis a decade before. This facility, the adviser added, offers a semiretirement package, too, in which younger residents can help provide companionship and care to fill some of the gaps—and earn a little money to offset costs. "They have a great system that matches up companions," the adviser said. "I have a handful of clients who really rave about it. I could put you in touch with one of them."

Jules liked the idea, and she knew Ava and Willow would appreciate that some algorithmic input would help smooth the transition. But she also knew the girls would rather they stayed home. Willow got over to China often enough, and Ava might consider moving to Europe. But would that be enough?

Either way, Jules knew they didn't have a ton of options, but as she and Gabriel walked back to the house that afternoon she felt grateful, knowing they could afford more choices than most people. Ava had sent them both a hologram, saying she felt a little better and thanking Jules for lunch, so they stopped at a café a few blocks from home to have a glass of wine, share a brownie, and decompress a little bit. Later that evening, they officially narrowed their decision down to the two choices they expected from the start: China or Germany.

"So, *ni hao* or *Guten Tag*?" Gabriel asked.

Jules chuckled. The certainty of only Option A or Option B had put her at ease. She knew which one she'd prefer, and she figured Gabriel probably leaned the same way. But she felt a the same dual twinge of excitement and doubt about both. And she could deal with that.

WILLOW (THE SISTER)

Willow was always the scientific wunderkind, winning her junior high science fair four years in a row, including the one when she was still in fifth grade and convinced her teacher to let her enter a machine-learning program that helped match her classmates with seventh-grade tutors. No one would've guessed that her eighth-grade project would eventually become the core of her groundbreaking environmental-preservation models, but they wouldn't have been surprised.

The problem with pure science, though, is that it doesn't pay. Willow open-sourced the Muir Models, named after her favorite national park, posting both the code and a white paper explaining the inspiration and a few initial notes for new directions the system could take. And with that, she decided, it was time to stop scraping by on loans from mom and dad. She called in a few favors from the grad students with whom she consulted, and soon fell in with an investment fund that needed an AI platform to help it balance better returns with the volatile political and cultural sensibilities of a global investor base. "If I can help them increase value by investing in the right kinds of companies, who cares if I'm selling out a little bit," she said as much to herself as to her sister and her boyfriend. Ava was skeptical, but she could see her sister had made her decision already.

Less than two years later, even Willow had misgivings. The advanced platform proved more successful than even she expected, but as they pushed the boundaries it became clear that, for all the fund's interest in sustainable investments, profit still came first. Twice, US regulators asked to review the code, and both times the company rejected their inquiries, citing the decade-old Algorithm Trade Secrets Act both times. The Chinese and Europeans demanded the code from top to bottom, so the fund never even bothered to register in either market. Willow wished they had, if for no other reason than to have someone confirm whether the platform went too far.

"You just want everyone to know how awesome your code is," her husband teased. Willow couldn't entirely deny it. She was proud of just how cleverly the program could tease out insights from almost any combination of relevant data sets. She'd already started tinkering at home with ways to apply some of the same techniques to her Muir Models, but knew the hedge fund would march her straight into court as soon as she released even the first line. She had all the material wealth she could want, but now wished she could hop back to science as easily as she hopped into the commercial world. Before he proposed, her husband had moved from Shanghai to San Francisco to head up UC Berkeley's computer science department—a prestigious enough post that even his company and the Chinese government helped smooth the transition. But standing in their Wyoming cabin, gazing past the gnarled ponderosa pine outside the picture window, Willow couldn't see anything but barriers in her way.

She stared past the tree and at the mountains for a while before sending the VR video to her mother. She always loved the home as a retreat, and began spending more time there lately, trying to figure out what to do. She'd expanded on her investment programs, privately tinkering around with ways to use AI to go beyond existing sensor systems and learn from different natural processes themselves. How does the forest solve pollution? How does the ocean convert oil spills? It wasn't that far off from the question the investment firm had asked—what, exactly, makes humans and machines decide they like a stock enough to buy it in a volatile marketplace? The profit-seeking motive followed the old reductionist tendencies of artificial intelligence, boiling complexity down to small sets of key variables. And it worked remarkably well. But Willow, always the curious one, didn't want simply good. She wanted to use AI to help embrace and revel in the greater complexities of the world—to absorb and process and enhance more of the messiness of life.

Her mother made no effort to disguise her displeasure when Willow went to work for the fund. So now, sitting at her desk in Wyoming, Willow still hoped a return to her environmental passions would soothe her mom. Willow's PAL chimed, and she nodded in return. "Ava is going to meet your mother for lunch. Her appointment with her doctor has concluded and she appears to be a little distressed. Would you like me to connect you?"

"No, I'll call later," Willow replied. "But cue Ava's PAL to play some Dave Brubeck at some point after lunch. She'll appreciate that." She waved her arm to dismiss any follow-up inquiries from the system and went downstairs. She needed a hike to clear her head.

It was nearly dark by the time she returned, and the house had transitioned itself to the evening. From the outside, with the chill descending off the mountain, their home looked warm and welcoming. She walked into the kitchen and started putting the leftover stew on the stove when her PAL chimed and patched Ava through. The immediacy jolted Willow after her quiet hike, but she could hear the urgency in Ava's voice: "I talked to Dad. I think they're going to move to Germany." Willow went numb. She knew Mom and Dad were thinking about moving overseas, but she always assumed they would figure out a way to stay home—even if it meant swallowing their pride and letting their daughters pitch in.

"They can't go to Germany, Ava," she said. "We'll hardly ever see them."

Ava gave a tired sigh. They'd gone over this before, and nothing had changed in their minds. Mom and Dad absolutely should stay at home. And if they didn't, well, Willow and Ava had different ideas about where they should go. But that they could deal with that. "Ava, tell me something," Willow said. "Do you think they'd change their minds if I told them I'm leaving my job to work on this natural complexity theory? I might take some of my code along, even if I get sued for it."

A short pause later: "Really? You're really going to do it? They might . . ."

EMILY (THE GIRLFRIEND) AND
AVA (THE PROTAGONIST)

She steeled herself to hear it again, trying her hardest to make sure she didn't show that the message phased her. "Thick skin; goddamned dinosaur hide," she told herself. And then she told her PAL to play it.

"I know a lot about you, Emily," the voice on the PAL said. "I know you don't believe a word you say on your show. I know you're dating another woman. And I know what you did to her ex-boyfriend. I also know that you have no idea what I'm going to do about it next."

"That's all there is. No number, nothing," Emily said, as confidently as she could muster.

She'd reported threatening callers before. Sometimes her managers offered support, sometimes they just mumbled something about "hanging tough" and criticism being "just part of the territory." She usually let them slide off her shoulders, venting to Ava when she got home. They both might fret about it for a day or two and then, together, let it all go. This one was different, far more personal and directly related to Ava. How would anyone in her audience know about Ava? Her bosses didn't even know about Ava.

A quick run of the metadata and voice through NewsHive's security platform flagged the message immediately, informing the local police and federal authorities. Emily saw the surprise on her manager's face when he saw the security system's note about the FBI. She'd resigned herself to the fact that this day would come, eventually, but not like this. She just assumed it would leak out, or some controversy would convince the company or a hacker to dig a little deeper into her data and start making the connections. An investigation into the caller would lead straight to Ava. A hard-right, conservative, Bible-thumping pundit who's actually a middle-left lesbian living in San Francisco? Yeah, that wouldn't go over very well with anyone.

The chaotic months that followed almost broke Emily. At work, her managers insisted they didn't care and wanted her to continue, at least through the end of her contract. They could begin building a new virtual character in case of emergency, but they wanted Emily to personally hand off the reins to a successor. She tipped ever more toward conspiracy theories, finding ways to support her arguments with data long debunked but still considered credible by her followers.

At home, Ava felt Emily swing further right and started to worry whether she was starting to believe the crap she spewed on her show. Ava confronted her almost every day now. Their relationship app—the one that used to just coordinate calendars, but now managed so much more of their personal interactions—kept nudging them into arguments, urging Ava to get things out in the open and prompting Emily to compartmentalize her work and home life. It all blew up seven months into the investigation, when Ava floated the idea of a party to remember Leo and said she wanted to send an invitation to Connor.

"It had to be Connor all along," Emily fumed. "Can't you see that this is exactly what he wants? He wants us fighting. He wants to get back into your life." She'd stormed out the door and Ava heard her slip about halfway down the stairs. Emily unleashed a string of venom on her way out the front door and onto the street. She didn't come back that night, and the next morning Ava sent out the alert to their shared network of friends, trying to keep it casual and not spark too much concern.

"Where the hell did you go?" she asked when Emily walked in.

"I went to see a few people," Emily said. "You don't want to know who."

Ava knew enough not to ask; Emily would never reveal those contacts. They sat in silence for almost an hour after Emily had showered, drinking cabernet and trying to figure out what had happened and how it could happen so fast. Ava finally broke the silence, asking her PAL to turn on the lamp and then looking at Emily. "This isn't Connor," she said. "He was really bitter when he left, but never about you."

"I know," Emily said, her jaw tight. "None of this made any sense, so my people hacked through all our shared systems to see if they could find anything. Turns out a vile little piece of code opened a couple of our PAL apps to a phantom data stream. Someone was watching all our information and feeding in fake data. Our apps did a pretty good job of recognizing the quirks, but this guy was pretty good, really subtle. Eventually, though, your Bouncer Bot started sending some fake streams to different places to see if it could identify the attacker. Someone was trying to drive us apart, but it wasn't Connor."

Emily was trembling now. Ava, sobbing, couldn't figure out if that made the invasion better or worse. "Is it fixed?" she asked.

"Yeah, my contacts cleared it out after setting a trap and turning everything over to the investigators," Emily said. "The FBI already had your metadata, and they'd been tracking this guy for months now. Their code forensics team got a few hints about him a while back, and then their artificial identity teams started tracking his digital footprints. Apparently, he loves to drink Woodford Reserve bourbon, and they used that to track him to some small town in Canada. They predicted where he'd come back across the border and nailed him. He's Russian, but he went to high school here in the United States."

The next day, as Ava walked back up the hill from work to home, the recollection made her knees buckle. Years ago, Connor introduced her to an old Russian school friend at one of his parties. She glanced around, relieved no one had seen her stumble. Her mind was racing. Her gut still told her Connor would never do it, but she had to know. She pulled out her PAL and told her Bouncer Bot about the memory.

"I'm aware of the connection," it replied, jarring Ava by speaking, as usual, in Connor's voice. "We're investigating a recent meeting between the two of them. You must know the probability that Connor did this, while quite low, is still there."

Ava got home and went straight to the half-empty bottle of cabernet on the counter. She reclined on the sofa, her mind slowly starting to ease with the hope that investigators would make an arrest and this whole thing would end. *Who knows what will happen with Emily*, she thought, *but either way I need to hear it from Connor himself.* She added him to the invitation list for the part to celebrate their lost friend, Leo. Something about the decision felt certain, and Ava laid her head back on the arm of the couch.

The lights had automatically dimmed and the speakers quietly sounded the hints of waves lapping the beach. She had already dozed off when the lock clicked open. Emily was home.

8

A World Worth Shaping

The morning arrived wet and cold, and the January 1215 day wouldn't get much better for King John of England. Having just returned from France, battle worn and financially strapped, he now faced angry barons in his own backyard. They wanted to end his unpopular *vis et voluntas* ("force and will") rule over the realm. So, to appease them and retain his throne, the king and the Archbishop of Canterbury brought twenty-five rebellious barons together in London to negotiate a "Charter of Liberties." The charter would enshrine a body of rights and serve as a check on the king's discretionary power.

The meeting that winter's day launched an arduous process, one fraught with tension and a struggle for power and moral authority. But, by that June, they had hammered out an agreement. It provided the barons greater transparency and representation in royal decision-making, limited taxes and feudal payments, and even established some limited rights for serfs. The famous Magna Carta emerged an imperfect document, teeming with special interest provisions and pitted against other, smaller charters, but it established one of the world's first foundations for human rights.

It would take another 300 years and multiple iterations for the Magna Carta to gain its stature as a reference for property rights, fair taxation, judicial processes, and a supreme law of government. Legally challenged throughout the centuries, the contentious document nonetheless prompted a dialogue about more democratic governance beyond England's borders. When settlers arrived on the shores of North America, they established their own charters for the colonies, followed eventually by the Constitution and its Bill of Rights, which brought to fruition the ideals seen in the Magna Carta and established them for every citizen, regardless of title and birth.

Today, we regard the Magna Carta less as a historical example of binding legal code and more as a watershed moment in humanity's advancement toward an equitable relationship between power and those subject to it. But it also marks the beginning of an unruly period in history, a transition that included the movement of people between continents and across oceans, the emergence of new political structures, and the many deadly conflicts that would arise between developing nation-states in an increasingly connected world. It set the stage for dialogue between powers, leading eventually to the Enlightenment, the Renaissance, and constitutional democracy, which rose out of much bloodshed over the course of centuries. Though mistakes were certainly made, it is probably also fair to say that—at least for Western countries—it provided fertile ground for the institutions and political and regulatory governance frameworks that eventually led to economic growth in trade and investment, as well as human and civilizational growth through the dialectic between powers in societies on the European and American continents.

As we delegate more judgment and decision-making to AI-based systems, we face some of the same types of questions about values, trust, and power over which King John and his barons haggled centuries ago. How will humans, across all levels of power and income, be engaged and represented? Which social problems, made solvable by powerful intelligent systems, should we prioritize? How will we govern this brave new world of machine meritocracy, so AI reflects the values and interests of society and leads to answers for humanity's most difficult questions? Moreover, how will we balance conflicting values, political regimes, and cultural norms across the world's diverse societies?

These questions take us well beyond technology and economics. Cognitive computer systems now influence almost every facet of life in most of the world's largest industrialized countries, and they're quickly pervading developing economies, as well. But we can't put the genie back in the bottle—nor should we try. As we note throughout this book, the benefits of artificial intelligence and its related advanced technologies can transform our lives for the better, leading us to new frontiers in human growth and development. We stand at the threshold of an evolutionary explosion of opportunity for the betterment of the human experience, unlike anything in the last millennium. And yet, explosions and revolutions are messy, murky, and fraught with ethical perils. We need to harness the vast potential of AI systems for human and economic growth, but to do so we need to consider the social ripple effects that cognitive systems, like the millions of bacteria and viruses in our lives, generate in our brains and our societies. Hence, as we set new applications out into the wild, we also need to put in place the key elements of an enabling infrastructure—the digitization of sociocultural norms with respect for their diversity, appropriate sensor and data collection technologies, governance institutions, and the policies and laws that will regulate development and deployment.

Because cognition pervades all areas of life, these structures can't be built by technical, commercial, or government experts alone, nor can they be prescribed in sweeping policy measures that don't comprehend the kaleidoscopic variety of AI applications and their potential. We need a range of historical, anthropological, sociological, and psychological expertise that can encompass the diversity of thinking about entire value systems in communities and how advanced technologies will influence them. We need to figure out which application areas raise which issues, and then neither over- nor under-restrict them if we want to realize their potential. And most of all, we need to make sure that the widest array of voices is heard and their values respected, so AI can be deployed for the common good of all humanity.

To accomplish this, we have proposed and have started working on a modern digital Magna Carta for the Global AI Economy—an inclusive, collectively developed charter that embodies the objectives and responsibilities that will guide the ongoing development of AI-related technologies and lay the groundwork for the future of human-machine coexistence. By integrating economic, social, and political contexts, this charter should begin to shape the collective values, power relationships, and levels of trust that we, as humans in this shared world, will expect of the systems and the people and organizations that control them.

We will need a living, breathing charter, one malleable enough to adjust to the inevitable disruptions that AI will generate. It should establish a global governance institution with a mutually accepted verification and enforcement capacity. We have dubbed it the "Cambrian Congress" to signify that the explosion of human growth opportunity could be akin to the explosion of life during the Cambrian Era in earth's history. The Congress should involve public, private, NGO, and academic institutions alike, with a collective aim of supporting human empowerment through agreed-upon norms. It should encourage the levels of transparency and explicability that foster greater trust and parity between human and machine intelligences. And it should foster a greater, symbio-intelligent relationship that enhances our humanity and its potential, not just machine efficiencies.

After all, the use of artificial intelligence and its cousin technologies can facilitate great advancement and terrible consequence in equal measure. Without a charter or congress to develop and oversee it, we might not mitigate the serious threats to our safety and security, but we also might miss out on an unprecedented opportunity to enhance our collective humanity and understanding.

BUILDING ON A RICH NETWORK OF EXISTING INITIATIVES

Sometimes, an idea just arrives before its time, and sometimes the innovations can't come fast enough. Sepp Hochreiter has seen both sides of that dilemma. Hochreiter is head of the Johannes Kepler University Linz Institute of Bioinformatics, but in AI circles he's known for his invention of "long short-term memory" (LSTM). The process essentially allows an AI system to remember certain information without having to store every little bit and byte of data it consumes. When Hochreiter and Jürgen Schmidhuber tried to publish the concept back in the mid-1990s, no one was interested. Conferences even rejected the research paper. Now, LSTM makes almost every kind of deep neural network more effective, providing a platform for feasible autonomous vehicles and vastly more efficient speech-processing applications.

While that idea sat in obscurity, Hochreiter says, he started applying his skills to the deep data needs of bioinformatics, where he could find plenty of work. Now, he's driving breakthroughs in pharmaceutical research. The potential effectiveness of any new drug must be weighed against its harmful effects, and toxic elements often lurk in the most minute nooks and crannies of the incredibly complex molecular structures that constitute today's drugs. Hochreiter and his colleagues have developed an AI model that can sniff out many of those toxic substructures. However, it didn't stop at the known toxicities, it started to discover new, often smaller structures that could produce similar ill effects—substructures previously unknown to the pharmaceutical researchers. "I have a neural network that knows more about the chemistry you're employing. It won't retire. It will work day and night. It won't leave your company. It will stay there," Hochreiter says. "But you have to turn data into knowledge you can make a decision on. Do I go with this compound or no? You have to transfer the data into knowledge, and from knowledge you make decisions."

We can take heart from Hochreiter's discoveries, both for the future of powerful new drugs that can cure disease, but also from the value of blending historical knowledge with our view for the future. Fortunately for the governance of artificial intelligence, we have a field of precursors from which to draw inspiration and initial guidance. It's heartening to see new governance initiatives and organizations emerge from commercial, professional, and academic sources, but few of them bring together all those varied interests in any meaningful way—none of them in a manner holistic enough to facilitate the broad agreement needed for an effective global charter. John C. Havens and the committees who developed IEEE's *Ethically Aligned Design* report might have produced the most comprehensive global representation of any such effort, having solicited input from developers, ethicists, and other interdisciplinary AI experts from around the world. The document will help establish standards for the technical aspects of AI development for IEEE members. What impact it might have on other spheres remains to be seen.

Others, such as Wendell Wallach, a lecturer at Yale and a senior adviser at the Hastings Center, have proposed ways to spread that thinking and a potential governance structure across broader spheres, including with "governance coordinating committees" that involve government, private industry, and civil society institutions.* Even some of the world's Digital Barons are looking ahead toward potential outcomes and how to prepare for them. The Partnership on AI is perhaps the most widely known initiative to emerge from the corporate sector, bringing together partners such as Apple, Amazon, Facebook, Google, Microsoft, and IBM, as well as the American Civil Liberties Union, Amnesty International, and other organizations. The Partnership also works with OpenAI, a nonprofit AI research company initially sponsored by Elon Musk, Peter Thiel, and others to find a pathway to safe AI development. In fact, many of the large efforts to emerge from the commercial sector work through a foundation or NGO structure, and often involve multidisciplinary participants, including links to policy makers.

Academically rooted institutions also bring together commercial and political interests, but rarely in a concerted effort to build actual governance structures. In England, some of the world's great thinkers have convened at the Cambridge Center for Existential Risk and the Future of Humanity Institute, both of which seek to develop strategies and protocols for a safe and beneficial future for advanced technologies. The Leverhulme Centre for the Future of Intelligence has crafted a sort of hub-and-spoke approach, assembling a team of top minds leading efforts around the world and forging connections with policy makers and technologists. Stanford University has launched a 100-year effort, called AI100, to study and anticipate the ongoing, high-level ripple effects of these technologies in all facets of life.

International and quasi-national organizations have jumped into the game as well. The United Nations Interregional Crime and Justice Research Institute has created the Centre for Artificial Intelligence and Robotics. In late 2016, the World Economic Forum founded its Center for the Fourth Industrial Revolution in San Francisco, hoping to create a trusted space where various public, private, and civil sector stakeholders could collectively cultivate the types of policy norms and partnerships needed to foster beneficial development of advanced science and technology. Hoping to help formulate some reliable,

---

* Wendell Wallach, and Gary Marchant, "An Agile Ethical/Legal Model for the International Governance of AI and Robotics," Association for the Advancement of Artificial Intelligence (2018).

human-centered governance standards that countries around the world might embrace—and using its convening power to help facilitate it—the WEF opened a second center in Japan in 2018 and plans to establish others in China, India, and as many as seven other countries by the middle of 2019. The WEF also established its Council on the Future of AI and Robotics, convening top government, corporate, and academic leaders from around the world to chart a roadmap for the centers' AI and robotics initiatives and to consider global governance structures for those technologies.

The WEF conceived of the center as a "do tank," running practical pilot projects to see what works rather than coming up with abstract theories that cannot be acted upon, says Kay Firth-Butterfield, the center's global AI Head. The projects are co-created with partner governments, businesses, civil organizations, and academics with a view to those partner governments piloting the policy and then adopting it afterward, she says. The center's AI-related teams have already started experimenting with a variety of novel ground-up approaches to technology governance. One such project established a depository where professors and other instructors could upload the curricula they use for teaching ethics and values within AI and computer science programs. By making those available to other teachers around the world, who can customize the instruction for their own situations and cultures and then share those, the initiative could develop global best practices and accommodate the world's diversity at the same time.

They also supported a remarkable on-the-ground program created by the Center's drone team, which worked with Rwanda as the partner government. That initiative used medical drones in the African country to help establish a standard for autonomous flight that several other nations have adopted in the months since. Rural women in Rwanda faced severe risks during childbirth because the country's medical personnel couldn't get blood supplies to remote areas if serious bleeding problems arose. So they started using drones to quickly send blood to clinics where it was needed. As of May 2018, a Silicon Valley firm called Zipline had delivered 7,000 units of blood on more than 5,000 autonomous flights across the country.[*] However,

---

[*]   Sarah Salinas, "The Most Important Delivery Breakthrough Since Amazon Prime," CNBC, May 22, 2018.

the Rwandan aviation authorities eventually objected to the unregulated use of drones in national airspace. So, the various parties came together, along with the WEF center, and developed the world's first performance-based regulations for drone traffic. Drone operators must meet certain safety and operational standards, and aviation regulators will account for qualified drone use in national airspace—potentially opening up Rwandan skies to a variety of innovators and their ideas. Regulators can specify certain safety standards, but then they also will accept qualifying drone missions and operations. "Now my colleagues on the drone team are getting requests from other countries to use that model to help commercialize drone deliveries," Firth-Butterfield says.

EFFECTIVENESS REQUIRES INCLUSIVENESS

These and so many other initiatives are advancing global awareness of AI values, power, and trust. Yet, we believe each of the current initiatives suffers from at least one of three imperfections. First, many are top-down or driven by a technological, societal, or political elite. No matter how well intentioned, a small, powerful subset of the population can only guide us so far in a field that's starting to pervade almost every aspect of our lives, often in very personal and identity-shaping ways. If we take one lesson from our past, it ought to be a concerted movement away from a winner-take-all exploitive power grab that has led to some of the darkest chapters of world history. Hard work remains to ensure a broadly accepted charter serves all humanity, something we might better accomplish by adopting elements of the WEF model or looking in unrelated fields—perhaps bringing in NGOs that have little to do with technology and everything to do with a deep understanding of human life patterns and the human condition broadly.

Second, none of the forums have an explicit mandate to assess and then facilitate social impact for the world's socioeconomically underprivileged classes. We might yet develop ways to use advanced technologies to bring more citizens of every standing into the decision-making process. Can we give them a voice in the cognitive power game to help define where we should focus our efforts?

Third, Western representation and thinking tends to dominate almost all the leading organizations to date. AI cuts across geographic boundaries and sectors of society, so a new charter must eventually actively solicit representation from all societies and their diverse interests—whether technical, philosophical, theological, socioeconomic, commercial, cultural, and so on. The voices of those outside the technology mainstream, such as rural residents in South America and Africa, need to be heard. These are the new economic growth centers from which locally and culturally appropriate innovation will spring, and their value systems can help inform global approaches, as the Rwandan drone example suggests.

As an inevitable first step in filling these three gaps, the government of each country should consider and explicitly state where it stands on the coming disruption and the voice its people should have in guiding it. Such an important threshold in human development cannot be left to chance, and only those with a clear point of view can contribute meaningfully to a collective future vision. This need not be strictly sequential, because a multistakeholder forum can inform national positions. But with national positions starting to form, we can start to build the international coordination and negotiation of a modern digital charter. Countries might even appoint AI ambassadors with clear lines to national leaders and to one another, providing direct links for quicker reactions to rapid changes and issues that emerge. (These networks might even include a few key private-sector representatives, given that the general user agreements on Facebook and other huge platforms already represent their own sort of social contract for a quasi-state.) The ambassadors could and should bring a range of interests and capabilities, including traditional industry skills, commercial interests, social and humanitarian concerns, and computer science expertise. They could bring the unique capabilities of their home countries to the global digital domain, helping broker solutions to problems in the most disrupted spaces of global society. Ideally, the principles that govern these efforts would be outlined in a widely adopted and mutually enforced accord, one embodied in a formal document with a supporting organization to coordinate, monitor, and enforce as necessary.

As a first order of business, the Congress would need to have a think tank function that is the diagnostic engine and prioritization funnel for the Congress. The staff could track and analyze global developments in AI, assess their impact on societal systems, and schedule deliberations about them in public plenary sessions.

To that end, it would work up case studies in priority areas of AI deployment, identifying the issues, second-order societal effects, and incentive schemes for actors involved in each case. This could prevent an overly broad dialogue that meanders around amorphous philosophical terms and doesn't lead to specific actions or responsibilities. (The last thing this world needs is more endless discussion without action or prescription, but we also need to avoid over-action and over-prescription that can kill innovation and hinder much-needed breakthroughs.)

This might sound similar to global governance organizations already in place, perhaps in entities such as the United Nations, the International Telecommunications Union, the World Trade Organization, and the International Monetary Fund. Yet, it differs in a material way. To embed good-faith collaboration on something so commercially dominated and so broadly pervasive in our lives, we need to directly involve and grant authority to private-sector representatives, such as the social entrepreneurs creating the world of tomorrow, governments of countries in which AI development lags, and the wide spectrum of NGOs that can represent disadvantaged or voiceless populations in different domains of social life. None of the institutions of the Bretton Woods system, nor individual NGOs such as ICANN, which manages domain names and IP addresses, have the proper design or the right competencies to deal with all the anthropological, legal, ethical, political, and economic dimensions these complex sociotechnical systems present.

It's a tricky balance to strike, though, as Terry Kramer's experience at the World Conference on Information Telecommunications suggests (see Chapter 5). The head of the conference, which sought to develop a set of globally accepted rules for the Internet, shut down much of the discussion that might have led to some common ground, Kramer says. Governments have a hard enough time trying to agree on major structures, and even less chance of agreement if they can't first establish a mutual understanding upon which to build, he says. When the dust had settled, the United States and about fifty-five other countries in the minority opted to not sign the WCIT treaty.

Of course, plenty of other agreements fall victim to the whims of politics and long-term neglect, as well. From the Paris Agreement to combat climate change to the Iran nuclear deal, from Brexit to the United

States' threats to pull out of NAFTA, all kinds of multinational treaties become more difficult to sustain in the face of new developments, disappointed expectations, eroded trust, and populist or isolationist rhetoric. By 2018, anxiety and anger about economic and political relationships went from a simmer to a full boil, with a growing cross section of the world population feeling like the global economy has been hijacked by a small, powerful elite, leaving everyone else with no input or voice.

A broadly inclusive, respectful, bottoms-up approach could help restore trust and, hopefully, avoid the fate of so many other global accords. But this, too, faces significant barriers—not the least of which are the more autocratic political regimes across much of the world, most notably in China but also across large parts of Africa, the Middle East, Latin America, and the Caucasus. Certainly, government representatives are indispensable, but governments alone lack the agility and expertise needed to facilitate profitable and responsible innovation. A coalition of private sector and civil society actors can track developments on the scientific and social fronts with much greater attention to important details, since they work where most of the functional expertise resides. All are needed in balance. The only way to avoid overregulation and the ceiling it can put on breakthrough innovation is to invite the private- and civil-sector actors into the fold. Conversely, the only way to establish effective authority is to keep government entities involved.

It might make sense to build a coalition through smaller, established government interactions and build from there. Already, Canadian and French officials are working to put AI on the agenda for the next G7 summit, building off the 2018 meeting in Canada and ramping up to the scheduled 2019 meeting in France. This might allow coordination across parallel tracks, getting buy-in from elite policy makers and corporate entities while also activating other, less powerful but equally important stakeholders. The United Kingdom, with its strong commercial and academic presence in AI development and its skepticism of the EU's data regulations, more closely resembles the US approach. A tighter philosophical alignment has begun to emerge among Berlin, Paris, Brussels, and, to some extent, Toronto and the Nordic countries as well. For people primarily concerned with a focus on the supremacy of humans, humanism, and democracy, the nascent alliance between these nations and the Vatican offers a more hopeful pathway forward. Whether the United States or the United Kingdom will join their ranks—and whether this tenuous coalition can muster enough global leverage to convince other democracies, much less other nondemocratic regimes, to join—remains to be seen. What's already clear is the global variation in political and philosophical perspectives about the roles of humans and machines.

As all these national governments join with large multinationals and other powerful entities to develop a working arrangement among themselves, they should also work with global foundations, NGOs, educational and scientific organizations, and small-business representatives to activate a broad range of vital stakeholders who might never get a seat at the table otherwise. In fact, much of this work can be led by global foundations, such as the Bosch, the Rockefeller, and similar foundations. These organizations already have deep insight into global society due to their extensive human-development programs, and they can use their considerable endowments to help invest in the computing power and other resources they need to impartially test programs and convene a progressive set of stakeholders.

NO EASY SOLUTION

How do we make progress in the wickedly complex environment of global artificial intelligence, which is so heavily dependent on values and has such high stakes? After all, even efforts to combat issues widely accepted as abhorrent often fail to reach optimum outcomes across the international community. The Organization for the Prohibition of Chemical Weapons (OPCW) has helped limit the development and deployment of such ordnance, but no accord is 100 percent effective. If the OPCW were, the international community might have recognized the development and prevented the use of chemical weapons in the Syrian civil war. If the Nuclear Proliferation Treaty had a perfect track record, we would not face current tensions over the burgeoning nuclear programs in Iran and North Korea.

Yet, in the domain of AI, two key factors give us hope for progress. First, as the world seeks to confront the complex challenges posed by a more fluid global trend, new or experimental forms of governance are emerging. While the nature of governance networks varies by context, in general we might be seeing a move away from traditional command-and-control models that attempt to enforce rigid and uniform rules.

New, more flexible models are emerging that promote greater participatory access, accommodate changing realities and situations as they emerge, and encourage a continuous dialogue between nations and other stakeholders. Because these new models facilitate an ongoing interaction across a broad range of participants, they seek to derive a more organic form of legitimacy and hold promise for the regulation of the rapidly changing and deeply pervasive field of AI.

Second, because of this multitude of players, the complex interplay between them leaves ample room for alliances both within and across borders. These partnerships can keep lines between countries from hardening and open up more creative space for finding solutions. Country, private sector, and NGO representatives can identify priorities and develop clusters of like minds to support them, creating beachheads for key issues that, even if not globally accepted, establish a foothold for certain values and goals. For example, on the jobs front, one might imagine an alliance between labor unions protecting their members; robotics firms concerned about a backlash; national political parties concerned about votes; the International Labor Organization upholding workers' rights; universities training students for the jobs of the future; and companies seeking to retain a productive and satisfied workforce. Such a cluster would lend itself to a holistic, design-thinking approach that could generate new solutions that rigid political stances might obstruct.

These models still raise questions about accountability and enforcement. Fortunately, because participatory regimes rely on the sharing of best practices and developing integrated solutions that work on the ground, they don't require the same hard-and-fast negotiating positions and the huge, monolithic stakeholders to sustain them. If one big country or corporation drops away, others can help fill the vacuum—although this works only if dropouts pay a price, whether in geopolitical interactions or in the global marketplace. One current example of this is the UN Global Compact, a voluntary corporate responsibility initiative that includes about 12,000 corporations and other institutions from about 145 countries. Its members accept a charter of ten guiding principles on human rights, labor, corruption, and the environment. The compact still faces a steady stream of criticism about accountability among its members, but the organization has started addressing some of those issues—including through one effort that invites representatives of harmed or weakened civil society groups to join and challenge violators in a public discourse about necessary changes. The initiative also includes a process that obliges members to report on their progress and, thus, opens the gates for public shaming when members fall short. Violators and laggards might find themselves reported to national institutions that have teeth, whether courts or regulatory agencies, which can formally audit and penalize them for misdeeds.

It's not an airtight solution in a world where Chinese, American, European, and other governments and companies have vastly different views on governance and regulation. No single solution exists. But even an assurance of an ongoing negotiating process carries weight. Rather than tackling one monolithic issue, competing interests might make inroads on partial issues and set up more sweeping agreements for the future. Already, mutual concerns have emerged worldwide about AI-driven cybersecurity, the safety of autonomous systems, and job losses due to automation. Common ground exists across all these issues, and many more, despite the sharp political and economic differences from country to country and company to company. Shared agreement can begin to shape the platform on which mechanisms for accountability and enforcement can stand.

A BROAD RANGE OF PERSPECTIVES ON TRUST, VALUES, AND POWER

The task ahead might feel Sisyphean, but let's not forget the social benefit that advanced technologies can provide. Even the most disadvantaged citizens can and should have a voice in their society and its values, and we can use AI technologies to empower them. Platforms on mobile phones could solicit and deliver feedback on proposed codes of conduct and their impact assessment—from the rural farmer in sub-Saharan Africa as quickly as the homemaker in Mumbai and the broker on Wall Street. The same AI-powered systems whose development we seek to govern can help in our quest, synthesizing millions of data points into a representative picture of value sets, human choices, and development potential. This doesn't happen immediately, of course—it starts with a diverse collection of existing interests—but any successful global initiative will need a clearly stated plan to gather the broadest collection of public sentiment, and a mechanism to ensure the integration of that feedback.

The effort to form a governance mechanism should solicit a range of views about how the integration of AI will influence the trust, values, and power that guides our lives:

- *How should we balance individual freedom of choice and societal interests in the use of AI? If you choose to opt out of an AI agent's recommended cancer-prevention program, should your insurance company know? Should it have the option to raise your rates?*
- *How should we deal with people who decide against the use of AI applications? If you refuse to sit through an AI-based video interview, what chances will you have to get the job you want, the job most suited to your talents or, potentially, any job at all?*
- *To what extent can AI support sociopolitical processes—such as elections, opinion formation, education, and upbringing—and how can we prevent harmful uses? If countries use social media as a battlefield, how will you know what is true, who is real, and whether your vote counts?*
- *How can we effectively counter the corruption of data sets and the potential discrimination against individuals or groups hidden in data sets? If cybersecurity systems can't protect your data, will your autonomous car stay on the road? And would a police officer treat you fairly if it didn't?*
- *To what extent should policies and guidelines define an AI system's respect for humans and nature? If an AI can solve food crises or climate change disasters, will you change your diet or your vacation plans to conform?*
- *How much importance should be placed upon social and societal benefits in the research, development, promotion, and evaluation of AI projects? If a computer or robot takes your job and the government pays you to do nothing, would you fight to stay employed? Would you look for a new job that gives you purpose?*
- *How can the promotion and training of employees for new employment and personal growth opportunities be integrated into AI-driven automation of production and work processes? If your employer assigned you an AI buddy that constantly nudged you to*

learn new tricks, experiment, and be more productive, would you stay or would you go?

- *How can effective, continuous exchange between different stakeholders be facilitated through AI? If you gave an AI a monthly budget and told it about your preferences, would you let it handle all your transactions?*

- *What type of permanent international institution would best facilitate the debate about and the governance of AI in seeking the greatest human and economic benefit? If an international group of people guided the AI agents that make decisions on your behalf, what would the group have to do to ensure your trust in those systems?*

Of course, neither the think tank nor the convening function of the Congress could take on all these questions at once. We need to focus on some of the highest value and most urgent issues—staying safe, keeping a job, and having a say over the direction of our lives. Each issue will need a leader, what Wendell Wallach and Gary Marchant call an "issue manager,"* who drives the debate and resolution across cultures. After all, these concerns cross borders, especially those on European and American continents, but across Russia, China, and countries with other political and cultural viewpoints, as well. A congress might have to start as a transatlantic endeavor, bringing together countries of like mind, and then bring in China, Russia, and the other countries we often spar with. It might be that China strikes out on its own, seeking to assert its bolder place in the world order. While not desirable, that could help forge parity between two primary geopolitical forces for AI development—a balance that might clarify both our differences and our commonalities and, perhaps, provide a platform for more constructive dialogue.

WHAT WE LEARN FROM THOSE WHO'VE COME BEFORE

---

* Wendell Wallach, and Gary E. Marchant, "An Agile Ethical/Legal Model for the International and National Governance of AI and Robotics," Association for the Advancement of Artificial Intelligence (2018).

The multinational treaties and governance models of our past might not provide cause for optimism, but they do offer guidance for mechanisms that assure trust, preserve values, and balance power. Getting everyone to agree to a common model is hard enough, but it might prove an even tougher battle to monitor AI development and enforce a new global standard of care. When we asked international governance experts and practitioners for examples of a comparable treaty that worked, most pointed to the same, single agreement: the Montreal Protocol on Substances that Deplete the Ozone Layer. The Montreal Protocol to reduce chlorofluorocarbons (CFCs) and mitigate depletion of the ozone layer was the most broadly accepted and most quickly implemented global treaty to tackle a tragedy of the commons problem, they said.

Jaan Tallinn came to the same conclusion after spending the better part of four years exploring these issues for a research paper on global cooperation. The cofounder of Skype more recently helped launch the Centre for the Study of Existential Risk and the Future of Life Institute, to which he contributes. His paper focuses first on humanity's experience with the tragedy of the commons—those instances when individuals act in their own self-interest and deplete, spoil, or ignore the necessary maintenance of a shared space or resource. His paper then explores the technological means we might have at our disposal to avoid such fates in the future, and what we might do with those tools as they develop. "There's no nation that has actively maximized steps to limit deforestation or global warming," Tallinn says. "It's a consequence of human activity. So, how do we all think two steps ahead of that?" The sorts of commercial, national, and economic incentives that drive the advancement of AI technologies don't exist for general AI governance. "The topic of AI safety seems to be a tragedy of the commons issue," Tallinn says.

So, what was it about the Montreal Protocol that worked so well? For one, it defined a shared problem—a hole in the ozone layer of the earth's atmosphere that diminished a protective shield against harmful radiation from the sun and increased rates of skin cancer and other unpleasant health impacts that nobody, regardless of political convictions, liked much. It then addressed this widely accepted problem with a tangible and easily understandable solution, one that had enough popular support to overcome industry objections. Next, it included provisions, including financial incentives, to assist countries as they phased out use of CFCs. Last but not least, it also had teeth, establishing certain trade actions against nations that refused to participate—the types of sanctions that, for a variety of reasons, haven't worked in many subsequent multinational negotiations. Global agreements to control the use of chemical weapons and landmines, govern the use of nuclear energy, and track the flow of small arms share many of those same attributes. Yet none of them brought together the same unique attributes as the Montreal Protocol—such as the sanctions or the broadly shared acceptance of a critical problem—and none saw quite the same level of success.

The Paris Agreement on climate change, signed in April 2016, provides some lessons, as well. Once again, participating nations faced a shared problem, and developed countries pledged to help developing states with pollution-mitigation investments. However, political support for the agreement varied across the globe, as environmental and economic interests clashed. The rift opened between countries that already attained the benefits of industrialization and those at an earlier stage of the process, creating what Todd Stern, President Obama's special envoy for climate change, called a "firewall division."* Unsurprisingly, the latter set of countries wanted the free industrial reins that the former group had enjoyed a century earlier.

However, unlike previous accords, including the Kyoto Protocol of 1997, the representatives in Paris established that any agreement needed to be applied uniformly across both developed and developing economies. It also let countries set their own emission-reduction strategies and goals, so long as those plans supported the joint mission of staying under the two degree global warming threshold. Critically, countries had to subject themselves to public scrutiny by submitting their targets ahead of the Paris talks. The combination of equal treatment and open oversight helped tear down the "firewall division," and it helped uncover certain allowances given to developing countries that richer nations could help address with private and public aid. As such, the Paris Agreement accomplished what few other accords had before: a bottom-up structure that not only facilitated a global commitment, but one that effectively addressed anxieties by remaining flexible enough to evolve and to allow different approaches in different

---

* Todd Stern, *Why the Paris Agreement works*, The Brookings Institution, June 7, 2017.

nations. Despite President Trump's announcement to withdraw the United States from the agreement, more than 175 nations remain active parties to it.

Unfortunately, artificial intelligence does not share some of the same solidifying factors that helped facilitate these successful climate agreements. Aside from the notable exception of AI's threat to jobs, most of its risks aren't immediately tangible. Few people understand how deeply cognitive computing influences their lives already. (The dystopian stories of robot overlords hardly help on this end. In that telling, how much of a threat does AI really present if it's not Skynet sending a muscle-bound Terminator back from the future?) Add to this the rapid globalization of trade and the shifting nature of multi-national competition and cooperation, and it's clear the guidance of our AI future must take a different shape—and derive its legitimacy from a combination of existing and new authoritative sources.

Patrick Cottrell studies the legitimacy and successes (or failures) of various international organizations, from the League of Nations to the International Olympic Committee. Cottrell spent the early part of his career in diplomacy, working at the US State Department before switching back to an academic track. He now teaches political science at Linfield College, where he wrote *The Evolution and Legitimacy of International Security Institutions*. Cottrell explains that most traditional intergovernmental entities, the United Nations in particular, derive much of their credibility through the "symbolic power" they radiate. "They stand for certain universalist principles, like peace and prosperity are good for all," he says. But these entities—born in a very different post-World War II environment and often formed by sovereign states—aren't necessarily well equipped to handle the governance challenges of the twenty-first century. Their design cannot easily adapt to transnational threats like forced migration, terrorism, climate change, and cybersecurity threats that cross borders and, especially in the case of cyber, operate in an entirely different dimension.

In response to these developments, Cottrell points to a growing, interdisciplinary body of work on global governance that is beginning to explore efforts to meet these challenges, particularly scholarship on "new governance." The speed and uncertainty of an AI-pervasive future call for the greater flexibility and wider participation of these models, which often operate separately from established governance bodies, such as the United Nations. As Cottrell notes, this approach recognizes the need for a broad cross-section of participants at the table, the proactive creation of new knowledge, and an iterative problem-solving design that recognizes that many successes will have to emerge from trial and error. "We can't possibly foresee the consequences of some of these things or anticipate them entirely," he says. "But we can create from a technology perspective, a policy perspective, and an industry perspective a mechanism that says, 'What sort of guidelines should we use to govern research, the ethics, [or] the dual-use application possibilities of AI?'"

If legitimacy is a social base of cooperation, as Cottrell argues, the legitimacy of this type of governance body would derive from its inclusion and the robustness of the norms and standards it disseminates. It may still work as an oversight body, but one with standards that evolve alongside new developments and mechanisms to disseminate best practices. It doesn't work if it doesn't adapt and generate standards from a wide range of participants. Still, Cottrell notes, an alignment with an existing pillar of global governance, such as the United Nations or the World Trade Organization, could help enhance the credibility of a new AI governing body. In this regard, a Cambrian Congress could nest alongside existing transnational organizations, where it can help crystalize shared values and build a consensus that national governments could use to negotiate more inclusive and successful treaties. And yet it could still accommodate individual national governments, which regulate their countries and are gatekeepers to civil action at home. It could also accommodate the necessary participation of an entire network of actors in AI—the separate entities that still share a common interest in governance of these advanced technologies. Only then can we guarantee that innovation opportunity doesn't get killed by preemptive fear, overly broad restrictions, or intergovernmental squabbling. "The UN Global Compact is perhaps one example that we might look to today," Cottrell says. The voluntary initiative calls on CEOs to commit to the United Nations' sustainability principles. It retains a clear tie to governments but "is very clearly shaped by corporate members, even though it's housed within the UN. That makes it a good bit more agile."

That sort of agility doesn't make it multistakeholder inclusive, but it matters, especially in digital technology, where transformative shifts in performance and scale materialize at a sprinter's pace. The slow, marathon speeds of traditional multinational governance could never keep up. Even a networked, solutions-based approach can't hit the same speeds, but it at least moves rapidly enough and remains flexible enough to maintain an appropriate threshold of monitoring capacity, accountability, and evolving mission in a rapidly changing AI ecosystem. Computer science experts can rotate in and out of different parts of the network, including the Cambrian Congress think tank, signing pledges and confidentiality agreements as they enter. Needless to say, compensation would be an issue because AI talent comes at a premium. But for this challenge, the AI sector might take some ideas from existing strategies. For example, Singapore's approach is to rotate its civil servants on a regular basis, enabling the public sector and its citizens to learn and share knowledge and skills across multiple industries. It still needs to attract talent by offering salaries that are competitive with the private sector, but with this approach it can maximize scarce resources.

GOOD PEOPLE FOR BAD POLITICS

Count René Haug among the skeptics. The former deputy head of the Swiss mission to the United Nations, Haug sees little evidence that governments would play together nicely. They certainly didn't when it came to the Organization for the Prohibition of Chemical Weapons, an initiative that one might think has rather broad support from the global populace. "If the OPCW is any indication, governments won't want certain information about the inner workings of smart technologies to be discussed and disseminated," says Haug, who now owns and runs a vineyard in Northern California. The implementation of a need-to-know system of monitoring, enforcement, and protection of confidential business information is expensive and all but impossible to carry out, even when widely adopted. Governments regularly squeeze through loopholes to sidestep monitoring, enforcement, and protection requirements, rendering confidentiality agreements effectively toothless. Even the OPCW suffers from that fate, Haug admits. For example, he says, the United States and other delegations disputed the need for the organization to install a strong encryption program for its servers, and the five permanent members of the UN Security Council basically requested unrestricted access to all the information compiled in those databases, including confidential business information. Countries also could request that their intelligence officers take staff positions at the OPCW, obliterating any chance the multilateral organization would have to safeguard private-sector confidentiality and competitiveness. The meddling of national interests proved essentially fatal, Haug says: "Lose that safeguard of confidential business information, and you lose the private sector."

UN programs to monitor and control the flow of small arms ran into similar challenges, says Edward Laurance, professor emeritus at the Middlebury Institute of International Studies in Monterey, California. Laurance and his peers who worked on the evolution of these programs over the years took three concrete steps that, in many ways, mimicked the elements that made the Montreal Protocol on CFCs so successful. First, he says, they developed enough data-backed evidence (the think tank function) to convince national leaders that small arms proliferation causes tangible societal and economic problems for all countries, not just those with the greatest amount of gun-induced violence. "Everybody started to see themselves as being in the game even if they didn't have the ball," Laurance says.

As a second step, working groups established voluntary standards and certifications for the trade of small arms, and then assigned networks of point people who verified compliance in their countries. They then pushed to establish a treaty—one that, at best, sees imperfect compliance but at least established a set of acceptable practices. "We ended up designing a bilateral certificate that was signed by the receiving or buying country and said, 'You must not sell these weapons across borders or use them on your own people,'" Laurance explains. That created a moral boundary and a framework for exposing non-compliant nations as clear outliers.

The last and, in some ways, the biggest hurdle is related to the talent and skills needed to carry out the agreements, Laurance says. UN and other government personnel might have the expertise to establish the processes needed for a treaty, but diplomats and agency administrators rarely have the level of technical expertise to carry it out. In the case of small arms, the involvement of military experts helped in many countries, Laurance says, because they know weapons, but monitoring and enforcement rules for advanced technologies would require an even more technically skilled cadre of experts. For that, private consultants and specialized academics would have to take part. "It's good to have that diversity and not rely on any one resource too much, because governments deemphasize an issue, skew science, or pull out of accords for political reasons," Laurance says. "Someone needs to stay" and provide the point of contact and a pool of expertise from each country and major stakeholder. Someone needs to keep the doors open for dialogue, debate, and small steps, even in tough times when bigger steps are hard.

The private sector and a strong coalition of NGOs might play an especially critical role in this regard. While governments and public policy can and often do change, broad participation of these private- and civil-sector entities might lend greater consistency and longer-term consensus. For example, given that many, if not most, American business leaders supported the goals of the Paris Agreement, President Trump's decision to withdraw from it might've been far less likely if those commercial and civic leaders were parties to negotiations, if not signatories to the accord itself. In a country where "big government" and political-intellectual elites are abhorred by many citizens, especially those who voted for Trump, influential private-sector institutions—unions, the American Chamber of Commerce, small and medium business associations, faith-based organizations, and top business leaders—might have swayed sentiment if they'd been stakeholders in the agreement.

As it stands today, though, few AI and advanced technology issues seem to capture the public's attention enough to present an obvious pathway to international governance. The one exception might be autonomous weapons. Lethal Autonomous Weapons (LAWs) capture the imagination as killer robots, but it includes far hazier territory than what Hollywood depicts in its sci-fi blockbusters. These military robots are designed to select and attack military targets without intervention by a human operator. Advances in visual recognition combined with the broad availability of small, cheap, fast, and low-power computing make the "autonomy" part of LAWs easy. A teenager could buy a computer motherboard for about $50 and program it to recognize faces. She could mount it on an $800 drone that could handle the weight. And, from there, it's not a great leap to the troubling realization that drone-based LAWs are plausible outside the control of military protocol.

In November 2017, UC Berkeley professor and AI expert Stuart Russell testified before the UN Convention on Certain Conventional Weapons in Geneva. He argued that autonomous weapons can easily become weapons of mass destruction because they're cheap to the point of being disposable, effective at finding specific targets, and, when deployed in arbitrarily large numbers, they can "take down half a city—the bad half." The video dramatization Russell shared as part of his testimony depicted micro-drones that fit in the palm of your hand, seeking targets via face recognition and killing them with small explosives.

The specter of small, inexpensive "slaughterbots" lends itself to shock and awe. The video quickly went viral, playing on the dystopian fears so many people hold in relation to artificial intelligence. But it's certainly worth the attention of citizens and governments. Several groups have been campaigning since 2015 for an outright ban on LAWs. As of this writing, US policy forbids LAWs to fire automatically, in part because it remains extremely difficult to capture in computer code all the factors that ought to go into determining who to kill when. More promising, though, the regulation of LAWs might eventually provide a useful hook on which we might hang some initial, broadly accepted governance. And perhaps that sets the stage for the tougher deliberations to follow.

THE MACHINE CAN MAKE US BETTER HUMANS

As these preceding global treaties suggest, we cannot build sufficient trust in artificial intelligence without the broadest possible engagement in the contentious debates that distill common values and strike an appropriate balance of power, especially as political and commercial forces ebb and flow. As a global society, we don't have a great track record of cooperation on governance issues, but we stand at a moment when such a collaboration might be more critical than ever. Not only do we need to mitigate the serious risks AI might pose to humanity, we need to capitalize on this unique chance to establish a fruitful ecosystem for advanced technologies, one in which powerful artificial intelligences enhance humanity and the world in ways we can't yet imagine. More than any other time in our history, we can curb the worst of human nature and capitalize on the best of humanity's creativity, imagination, and sense of discovery. Given our collective global capabilities, from scientific examination to entrepreneurial zeal, we can expand our frontiers and unleash our potential in this emerging cognitive revolution.

We face daunting obstacles, many of which challenge what it means to be human. The very idea of a powerful intelligence that mimics but, in certain ways, surpasses ours both fascinates and scares us. We gauge it against ourselves to determine whether it's a threat, and most people in Western countries view it as exactly that. This alone could keep us from tapping its full power and building a fruitful symbiotic partnership between human, artificial, and other intelligences. Our human spirit and our purpose carried us to heights unattained by other lifeforms on earth, with greater intellect and greater power. But we have not always used that advantage responsibly. When we do, and when we employ the technologies we create, we accomplish amazing things—500 million people lifted out of poverty over the last century, an additional twenty years on the average lifespan, and humans walking on the surface of the moon (and maybe, soon, on Mars as well).

What drives us to these peaks? We aspire to them, wanting our lives to matter and wanting to achieve for ourselves and for our children and grandchildren. Some of us interpret these things in totalitarian or aggressive ways, but most of us don't. We strive and we struggle, relying on each other to advance human and environmental well-being in small and large ways. Most of us don't leave others behind. We might forge ahead and sail to new shores, and our actions might seem egotistical and self-indulgent at times, but we usually return to show others the way across the expanse. Sometimes it's the entrepreneurs with the thickest skulls who produce the greatest impacts on our lives. Sometimes it's the awkward or antisocial scientists who chart the way to the moon, map the ocean floors, and scale the proverbial mountaintops on their insatiable journey to the frontiers of our universe, our minds, and our souls. Great human thinkers don't rest unless we advance and grow, and we grow in partnership with the people around us.

We thrive on empathy, imagination, and creativity—attributes rare or nonexistent in even the most powerful AI systems, yet abundant in humans whatever their nationality, ethnicity, socioeconomic class, or education levels. Our thoughts collide and spark ideas in a mesmerizing display of creative friction. And yet, every time we move closer to another human being to share a new thought or a new emotion, we blend that creativity with the depth of imagination and empathy that allows us to collaborate, innovate, dream, love, and build.

We always build, and we always destroy. We create the world of the next moment, the next month, the next year, the next century. We construct cold, hard tools to forge warm social and emotional bonds. We build houses and families to live in them. We build roads to connect our neighborhoods, and we build the types of relationships that make us neighbors. We create new technologies and economies, connecting across the globe to expand wealth and knowledge for tomorrow's generations.

We built globalization based on the Anglo-American principle of the free flow of goods, services, and capital, and it led to imbalances that were hard to take for many societies. Now we are deconstructing that model, even as a new type of globalization emerges—one that features a global data economy spearheaded not by Wall Street and the City (i.e., London as a financial center), but by the entrepreneurs of the digital realm. That too will lead to missteps and crises. Over time, we will tear down some elements and forge ahead with others. But make no mistake, the global data economy with its autonomous machines is here to stay.

That's because the new machines can build, too. They can create models of the future, often in better and more insightful ways than humans can. They can generate insights out of massively complex data sets and the subtle patterns that human brains can't process. AI systems already create alongside us, but they do so differently. They can't share the experience of human existence and the millions of years of biological evolution that's encoded in our brains and that serves as a base memory for the human condition. This shared experience forges a certain kind of common bond that machines can't share. Many of us express those connections through various religious or philosophical beliefs, almost all of which strive for righteousness, justice, care, and love. Do unto others as you would have them do unto you. Cognitive machines can and should emulate all these things. They can copy and empower us. But they can't *be* us, because they have never crawled out of primordial slime through the synthesis of cellular trial and error, success and failure, joy and pain, satisfaction and frustration, to stand on top of Mount Everest or on the surface of the moon, raising their arms and crying out to all humanity. Their learning is not reinforced by emotion. Their bodies are not designed to feed the mind with sensations from the myriad internal and external sensations that enhance a human's ability to relate to the environment and to survive and evolve within it.

As we forge deeper symbio-intelligent relationships, we can't forget that humans are unique—not necessarily better or worse, but unique nonetheless. Yes, we are one species among many, and we may need to grapple with the possibility that machines could evolve into another species alongside us. But as we do that, we also need to avoid the urge to think of the brain as little more than a computer analog, as Alan Jasanoff writes in his recent book *The Biological Mind*. The "cerebral mystique" belies the complex interplay between physical sensations, emotions, and cognition that comprises the whole of a human being, Jasanoff writes.* Someday, we might replicate human brains on silicon or some other substrate, but human cognition still requires a complex array of interactions with our bodies and environments. Human ingenuity might one day come up with a "replicant" that connects an intelligent brain to a humanoid body and taps the many stimuli of the physical environment around it. But until then—and that's a far, far cry from the level of artificial intelligence development we have today—machines will remain socially and spiritually stunted.

And that's OK, for we must also avoid the trap of equating human and machine, pitching the two as rivals in an existential race to the top of the intellectual pyramid. The complementary power of machine and human intelligence working together offers far too much promise to give way to an unduly competitive mindset. Our "wetware" brains are remarkably inefficient in certain cognitive methods—messy and bubbly and wrangled, easily distracted and prone to meandering. And yet they're wonderfully experimental. We transpose concepts from one field to another, although we might often do this in ways that have our friends furrowing their brows or rolling their eyes. We waste millions of instructions per second (MIPS) by staring into a sunset or watching the grass grow, making wild connections to far-off concepts that spark near-genius moments. Seemingly random blips in our mental constitutions have us bursting out into laughter, anchoring some crazy thought

---

*   Alan Jasanoff, *The Biological Mind* (New York: Basic Books, 2018).

with an emotion of delight that won't reveal its funky self until years later, when the most random occurrence provokes it.

Cognitive machines with their neural networking power can help us become more effective sensors and faster processors, so we strike more connections and make better decisions. They can help us build greater scientific constructs, human communities, or ecological resiliencies. And they can help us strategize by simulating incredibly complicated situations and running outcome scenarios, enhancing our ability to choose the best paths forward. We will need AI agents to help solve our worst and toughest problems—climate change, health care, peaceful coexistence—but they will need our vision, our spirit, our purpose, our inspiration, our humor, and our imagination. Paired with AI's analytical and diagnostic power, we can soar to new heights, understand nature more deeply and holistically, explore farther into the cosmos, and establish our place in the universe in more satisfying ways than we ever experienced before. Perhaps for the first time, we might achieve our aspirations for humanity while increasing environmental stewardship, stepping out of a zero-sum game and optimizing across so many more of the variables in our everyday existence.

Combining the unique contributions of these sensing, feeling, and thinking beings we call human with the sheer cognitive power of the artificially intelligent machine will create a symbio-intelligent partnership with the potential to lift us and the world to new heights. Some things will go wrong. We will second-guess ourselves at almost every turn. The pendulum will swing between great optimism and deep concern. We will take two steps forward and one step back, many times over.

But that is why it is so essential that we engage now and begin the dialogue and debate that, out of our rich human diversity, will establish a common ground of trust, values, and power. This is the foundation upon which we build the future of humanity in a world of thinking machines.

AFTERWORD

by Laura D. Tyson and John Zysman

I ntelligent tools and systems are diffusing through economies and societies around the globe, affecting how we work, earn, learn, and live. Our daily lives are already powerfully shaped by digital platforms such as Amazon, on which we buy goods and services; Facebook, through which we track our friends, even as we are tracked; and Google, through which we access a world of information. And the press is replete with tales of automated factories run by robots.

Accelerating the growth in the power of platforms and automated systems, as *Solomon's Code* makes clear, is the emerging flood of artificial intelligence tools. The functionalities and applications of these tools are diverse, but as Groth and Nitzberg observe, "At their core, all the various types of AI technologies share a common goal—to procure, process, and learn from data, the exponential growth of which enables increasingly powerful AI breakthroughs."

At least in theory, in the long run AI systems with advanced capacities for reasoning and abstraction could perform all human intellectual tasks at or above the human level. Groth and Nitzberg refer to this state as "Artificial General Intelligence." Others refer to it as the Singularity. Fears about the possible domination of humans by machines embodying artificial general intelligence are stoked by news stories, fiction, and movies. The specter of artificial general intelligence is raising profound questions about what it means to be human.

Whether, if ever, we arrive at artificial general intelligence, narrow AI tools that imitate human intelligence in specific applications are developing rapidly, resulting in what the press calls the "appearance of intelligent behavior in machines and systems." As Groth and Nitzberg's deep dive into the sweep of these new tools makes clear, there is an array of possibilities to capture as well as myriad challenges and concerns to address.

Many of the impacts of such tools are already evident. Consider the ability of platforms like Facebook to target advertisements and information to particular groups, even individuals; to affect political discourse and outcomes; and to provide new ways for people around the globe to communicate. We have information at our fingertips and remarkable capacities to communicate, but the information about us is widely available and the capacities to communicate often absorb a disproportionate amount of our time. Suddenly, not just a war of information and misinformation is evident. There are also disturbing signs of the makings of a surveillance society, as well as evidence that even the smartest algorithms might systemize rather than counter human biases and flaws in judgment.

Intelligent tools and systems are spreading rapidly, their power continuously expanding with the growth of AI tools, transforming how goods and services are created, produced, and distributed. Companies will need to adjust their processes, products, and services to the new technological possibilities to sustain or gain competitive advantage. But will the resulting benefits accrue to their work forces? Will we see massive increases in productivity as our societies become increasingly rich, but increasing inequality as the gains are shared ever more unequally? Will we see rapid displacement of work and workers, technological unemployment, and mounting inequality within societies and between economies?

Optimists proclaim that the future is ours to create. Easy to say, but the difficulty is that there is great uncertainty about the possibilities and challenges in a world of increasingly sophisticated AI tools and applications. Consider the impact of AI-driven automation on work and jobs, which is the focus of an interdisciplinary faculty group at UC Berkeley called Work in an Era of Intelligent Tools and Systems (WITS.berkeley.edu). There is broad agreement in research by McKinsey Global Institute,[*] the OECD,[†] the World Economic Forum,[†,§] and individual scholars including Kenney and Zysman,[¶] all finding that some work will be eliminated, other work will be created, and most work—as well as the terms of market competition among firms—will be transformed. There is also broad agreement that intelligent tools and systems will not result in technological unemployment—the number of new jobs created will offset the number of old jobs destroyed—but the new jobs will differ from those that are displaced in terms of skills, occupations, and wages. Moreover, it appears likely that automation will continue to be skill-biased, with the greatest risk of technological displacement and job loss falling on low-skill workers. A critical question, then, is how the new tasks and jobs enabled by intelligent tools and systems will affect the quality of jobs. Even if most workers remain employed, will their jobs support their livelihoods.?

Although there is widespread agreement that AI-enabled automation will cause significant dislocation in tasks and jobs, there is considerable uncertainty and debate about the magnitude and timing of such changes. Many routine tasks will be displaced or altered; but how many jobs will be displaced entirely? Will long-distance truck drivers become a thing of the past? Or in a future enabled by AI and driverless cars, will truck drivers be able to perform critical management tasks at the beginning and end of their journeys, sleep on the job as their

---

[*]   James Manyika, et al. *Jobs lost, jobs gained: What the future of work will mean for jobs, skills, and wages*, McKinsey Global Institute, November 2017.

[†]   Ljubica Nedelkoska and Glenda Quintini, "Automation, skills use and training," *OECD Social, Employment and Migration Working Papers, No. 202*, (OECD Publishing, Paris: 2018).

[‡]   *Towards a Reskilling Revolution: A Future of Jobs for All*, World Economic Forum, January 2018.

[§]   *Eight Futures of Work: Scenarios and their Implications*, World Economic Forum, January 2018.

[¶]   Martin Kenney and John Zysman, "The Rise of the Platform Economy," *Issues in Science and Technology*, 32 (3): 61-69, 2016.

trucks move across structured highways, and make deliveries along the way? Some Japanese firms, confronted with outright skill shortages, are already turning to AI, machine learning, and digital platform systems to permit less experienced, not less skilled, workers to take on more difficult tasks. More generally, the performance of routine tasks currently performed by humans might be converted into human tasks to monitor and assess routine functions performed by machines. What sort of new jobs or tasks will be created? What are the skill and wage differences between the jobs that disappear and the jobs that are created? Whether digital platforms, such as Uber, create a flood of gig work or new sorts of transportation firms with imaginative new arrangements for work will depend as much on the politics of labor and on labor market laws and rules—features that differ dramatically across nations—as on the technologies themselves.

For firms, communities, economies, and societies, adjusting to new transformational technologies is never simple. One need only look at the economic and social upheavals during the first industrial revolution. Crucial questions driving our research are whether and how technological trajectories supporting high-quality employment and skilled work can be created and sustained. Can we do so quickly enough to match or ideally be ahead of the high pace of this technological change that is already upon us? How do we incentivize corporations and civil society groups to engage in the definition of future skill sets and jobs? These are also the questions raised in *Solomon's Code*: How can AI-enabled technologies be shaped to support human purpose and well-being across diverse countries and regions with different political and cultural objectives.

The basic security of digital systems themselves also poses major economic and societal challenges. For firms seeking competitive advantage from digital AI tools, securing their financial, intellectual and operational assets is crucial. Cybersecurity has already risen to the top of business risks, necessitating significant investment in systems to prevent cyberattacks by competitors, individuals, or governments. Consider the energy industry. For communities trying to introduce renewable energy sources into the grid, digital control systems are essential but vulnerable to disruption by cyber hacks. Indeed, as is emphasized in the foreword by Admiral James G. Stavridis, such industrial or sectoral vulnerabilities can quickly become potential strategic vulnerabilities, as Estonia has discovered. Cybersecurity at the level of firms and institutions suddenly blends into concerns about cyber conflict, both between nations and as tools of attack for non-state actors.

The public policy challenges facing governments are diverse, difficult, and often conflicting. *Solomon's Code* gives us a good picture of the different ways several countries are proceeding. The *first* challenge is how to capture the possibilities that AI tools represent. However, this only opens the policy debate. Does policy for an AI adapted society call for a broad investment in the development of the tools themselves, as China is quite evidently doing, and which Israel pursues as part of its basic defense strategy? Does it call for policies to support diffusion and adoption throughout the economy, as arguably the Germans are doing with their program of *Industrie 4.0?* Does it call for investment in and the redesign of education and training? Or should governments simply stand back and encourage AI-driven disruption and dislocation, following an Uber like mantra of "don't ask permission, ask forgiveness," and deal with the mess after the fact?

To adapt to changing technological opportunities and challenges, existing institutions, from schools to justice systems to defense systems, will have to evolve. Apart from the difficulty of change within existing institutions, the appropriate objective end points are far from evident. Education is a case in point. Is it important for students to learn to code? Or is it more important, or at least equally important, to develop students' human skills—like empathy and creativity—alongside their ability to use and apply digital tools as they evolve. Do we need deep reform in secondary schools, or, assuming students are taught to read, write, reason, and code, will a series of nano-qualifications come to substitute for parts of higher education as currently organized? If the latter proves to be the case, are public systems or for-profit institutions the appropriate providers of the necessary skills? Redesigning an education system to meet the needs of a new technological era is much harder than simply expanding an existing system to involve more students and longer years of schooling.

A *second* set of challenges is protecting against the vulnerabilities and possible abuses in a digital system with powerful AI tools. Here, the question of what is private cyber-security and what is public responsibility in cyber conflict/war is critical. Who is responsible for the integrity of voting systems, of news, of financial systems, and energy systems? Societies like Finland do not separate these matters as sharply as the United States currently does.

A *third* set of challenges is the governance of the digital/AI system itself. Here, the array of challenges is as broad as it is daunting. Our basic rules about market competition, competition policy in Europe, and antitrust policy in the US, for example, are awkwardly adapted to a world of dominant platform players with global capacities to shape how we see the world and control vast amounts of information and data. How is market power measured when the products and services of dominant firms are "freely" offered to consumers in exchange for valuable information about them? How do we enforce anti-discrimination policies if biased judgments are hidden inside of non-transparent AI engines? What international rules are needed to protect privacy and to counter biases, and in what institutions should such rules be made?

A *final* set of challenges—and in our view, both an ethical imperative and a practical necessity—is providing social protections and capacities for individuals and communities to adapt and adjust to the dramatic changes and dislocations caused by AI-driven technological change. Providing transitional help to those whose lives are disrupted by technological change will not be enough to ensure social, economic, and political stability. Creating opportunities for renewal and new livelihoods—generating the narratives of the future for those who are displaced—will be essential.

Solomon's Code contains a strong and clear message. AI tools and the systems and platforms they enable are advancing more quickly than anticipated, powered by breakthroughs in computing power, semiconductors, and data storage. These advances are challenging individuals, businesses and nations to respond rapidly. Those that effectively adopt and adapt will find advantage. Those that either do not adopt, or do not adapt, are likely to be left behind. For nations, failure to develop policies to maintain equitable outcomes will be as disruptive as failure to adopt. *Solomon's Code's* rich mix of analysis and examples across a broad array of questions and issues that are raised by the deployment of AI-powered intelligent tools and systems provides an excellent starting point for confronting the choices and debates that lie ahead.

Acknowledgments

The cover of this book features two authors, but the list of people who made this endeavor possible runs far, far longer. No one sacrificed more than our wives, Dr. Ann Reidy and Prof. Elisabeth Krimmer, who patiently allowed us the time and support—sometimes even during family vacations—to research and complete this book. None of this would have been possible without their generosity and love, and we owe both our deepest gratitude for this and so much more.

As newcomers to mass-market book publishing, we had no choice but to rely on a deep bench of incredible talent. This project probably would have died an early death had it not been for the vision of Esmond Harmsworth, our agent at Aevitas Creative, who saw how the kernel of our concept could stand out from the existing crowd of AI-related material. Jessica Case, our editor at Pegasus Books, recognized the same potential, shepherding us through the development of the manuscript and helping sharpen it into the version you see today. Her team at Pegasus, including copy editor Meaghan O'Brien and proofreader Mary Hern, saved us from more than our fair share of embarrassing mistakes. If any errors remain, the blame begins and ends with us.

Long before these brilliant professionals worked their magic on our proposal and manuscript, a string of friends and consultants helped us refine our ideas and navigate our way into the publishing channel. Our friends Chris and Alexandra Ballard, successful serial authors, gave us an early glimpse of just how much would be required of authors who set out to write a mass-market book on such a technical and rapidly evolving topic. And, perhaps equally important, they introduced us to Dan Zehr, a writer (and now editor in chief at Cambrian.ai) who helped transform our style from one appropriate for professionals to prose accessible to a wider audience. Dan proved to be an invaluable collaborator, responsive to our ideas and adaptive to our turns and pivots, and gentle and patient at that. Over time, our collaboration arrangement morphed into a tight trio of friends who, during many writing retreats, not only got a lot done in 18 months but had a lot of fun at that.

Many others helped guide us into and through the publishing process, and we wanted to thank a key few here: best-selling author Michael Lewis offered a load of little insights that helped focus our direction and our writing process; Marilyn Haft's legal expertise allowed us to focus on reporting and writing rather than contracts; Jeff Leeson helped us understand the market for proposals and how to make our pitch stand out; and Dr. John Beck (Phoenix), Dr. Bhaskar Chakravorti (Boston), author Brian Christian (Berkeley), Prof. Mark Esposito (Lausanne), Rehan Khan (Dubai), Joanne Lawrence (Boston), and Jon Teckman (London) provided friendly advice and examples of their own creative processes from a diverse set of global perspectives.

The list of contributors who shared their remarkable knowledge with us is too long to mention in this space. You will see their names throughout the book; please consider each mention of each person a "thank you" from us. We do want to extend a special thanks, however, to a few contributors who don't appear in the book but provided further thoughtful commentary on AI in discussion roundtables and personal meetings that coincided with the research for the book: German Federal President Frank Walter Steinmeier; State Minister Dorothee Bär at the German Chancellor's Office; Cedric O, the Counselor for

Public Participation and the Digital Economy, and Thierry Coulhon, the Counselor for Science and Innovation at the Presidency of the Republic/Prime Minister of France at Elysee Palace; Dr. Jean- Philippe Bourgoin, Senior Adviser for Research at the French Ministry for Education, Research and Innovation; State Secretary Walter Lindner, Ambassador Peter Wittig, and Consul General Hans-Ulrich Südbeck at the German Ministry of Foreign Affairs and their ministry colleagues Ambassador Wolfgang Dold, Maria Gosse, Katrin aus dem Siepen, and Vito Cecere; as well as Gisela Philipsenburg and Katharina Erbe at the Innovation Policy Issues division at the German Ministry for Education & Science, Michael Schönstein, Deputy Head of Division at the German Ministry of Labor and Social Affairs, and Max Neufeind, Adviser to the German Minister of Finance and Vice Chancellor.

A host of multipliers helped us tap into new networks and gather perspectives in places we might never have experienced otherwise. We mention some of them throughout the book, but several deserve a special thanks here. Without the dedication of Hsiao-Wuen Hon, Jane Ma, and the team at Microsoft Research Asia, we could not have connected with the Chinese AI ecosystem to the extent required of a book like this one. Wen Gao, Yizhou Wang, and Mrs. Cai at Peking University helped build on our initial connections there, and James Chou, the CEO of the Microsoft Accelerate Shanghai, provided invaluable guidance as we researched the technological landscape there. Horst Teltschik, the former national security adviser to German chancellor Helmut Kohl, president of Boeing Germany, and board member at BMW, Thomas Neubert at Intel, Angela Chan at Channel 4, UK, Gerard Sheehan and Dorothy Orszulak at the Fletcher School at Tufts University, and David Reidy at the University of Tennessee also helped connect us with key insights, resources, and people who helped provide a most holistic picture of AI's influence on individuals and societies worldwide.

We set out to craft a book that would balance technical sophistication and appealing narrative, and much of the credit for any such success we achieved should go to our friends and colleagues who agreed to critically review various sections of the book. They include: Padideh Ala'I, Alexandra Ballard, Chris Ballard, John Beck, Thomas Bjelkeman-Pettersson, Amy Celico, Patrick Cottrell, Sarah Courteau, John Fargis, Kaolin Fire, Pascale Fung, Prof. Ken Goldberg, Owen

Good, Rehan Khan, Terry Kramer, Joanne Lawrence, Bertrand Moulet, Moira Muldoon, Erik Peterson, Dr. Ann Reidy, Thomas Sanderson, Crispin Sartwell, Peter Stone, and Elizabeth Zaborowska. Still others took time to peruse the entire manuscript and provide feedback and reviews. We can't thank them enough for their time and generosity: Beth Comstock, Brad Davis, Bill Draper, Prof. Mark Esposito, Dr. Evelyn Farkas, Ken Goldberg, Timothy Koogle, Erik Peterson, Arati Prabhakar, Lord David Puttnam, Peter Schwartz, James Stavridis, Shashi Tharoor, Prof. Laura D'Andrea Tyson, Jim Whitehurst, and Prof. John Zysman.

Importantly, we could not have built our network and crafted our international perspectives without the generous help of former students at Hult International Business School the Fletcher School at Tufts University and the Computer Science department at the University of California Berkeley. In particular, Aleksandra Kozielska, Mike Kuznetsov, and Bruna Silva delivered invaluable contributions in the form of global insights and marketing expertise. Massiel Acuna, Juliane Frömmter, and Vincenzo Ottiero helped open doors around the world, finding novel AI applications and helping shape our understanding about both major and overlooked markets. And a global network of students at Hult helped widen our field of view: Kailash Bafna, Line Boisen Petersen, Adrian Cevallos, Maria Dolgusheva, Tracy Katrina Ebanks, Laia Esteban, Eleonora Ferrero, Rodrigo Goulart, Robert Grüner, Francisco Guerra, Jasmin Jessenk, Joseph Kong, Jonathan Kurniawan, Bridget Lekota, Arjun Manohar, Maximilien Meilleur, Olga Matican, Hugo Anas Mekaoui, Anna Molinero, Alexander Neukam, Babatunde Olaniran, Olga (Tansil) Palma, Anna Podolskaya, Aaron Salamon, Suhail Shersad, Jayvijay Singh, Sukhdeep Singh, Tobias Straube, Chang-Hung Tsai, George Wang, Tatum Wheeler, and Esau White.

These students represent an emerging generation of critically reflective and responsibility-minded business and societal leaders. As we head into a new era marked by deeper human-machine collaboration, their passion, energy, and thoughtfulness constantly reinvigorate our hope for the future.

—Olaf Groth and Mark Nitzberg

Postscript

hen we initially published this book as *Solomon's Code* in
2018, we knew enough to expect some significant changes
in the field of artificial intelligence and the world beyond.
What we couldn't anticipate back then was just how disruptive the
next three years would be. The usual turnover of people going on to
new jobs, companies rising and falling, and new technologies emerging
to change the innovation landscape seem downright quaint when
compared to the upheaval of the global Covid-19 pandemic, the
Trump presidency, and a growing global backlash against the large
technology platforms and their algorithms.

Yet, all these changes have made the lessons of this book *even more
relevant and urgent than ever before.* The deployment of AI systems
across an array of sectors progressed faster than even we anticipated,
especially once the pandemic squeezed a decade of transition from
"partially digital" to "fully digital" into one extraordinary year.
Whereas the arguments in this book previously resided within the
domain of tech-savvy readers, the awareness of AI, its opportunities and especially its risks, has awakened mainstream public attention. Virtually everyone now has an inkling—and often a strong

opinion—about how the examples and lessons presented here touch their lives. Even if advanced professionals and ruling elites can't claim the label, the world's emerging digital- and AI-native workforce can truly call itself the "AI Generation."

Like any force that sparks cultural, societal, and economic upheaval, the central role of algorithms in our lives and our minds cuts in both good and bad directions, but today we've seen remarkable advances in recent years. Decentralized FinTech structures have started to form the foundations for a more efficient and more equitable financial system, coalescing amidst the volatile boom-bust cycle of crypto technologies. The powerful bio-engineering and medicine research capabilities we'd previously anticipated have come to fruition, as the confluence of AI systems and CRISPR technologies allows experts to find cures for obscure ailments faster. Combine that sort of innovation with unique, multi-company and cross-border collaborations, and you get examples like Pfizer and BioNTech working together to develop a Covid-19 vaccine in a remarkably short period of time. Meanwhile, two leading brain-computer interface (BCI) ventures—Elon Musk's Neuralink and Kernel, which is headed by Bryan Johnson—now appear to be about two years from commercial trials, putting them about two years ahead of the timeline we'd previously projected.

Advances within the field of AI itself have accelerated and generated their own well-deserved headlines, too. Although we still don't see any indication that artificial general intelligence (AGI) will develop in our lifetimes, narrow applications of AI have produced remarkable breakthroughs. Just two months before we wrote this, computer science pioneer Geoffrey Hinton published a paper about a new concept he called GLOM, which envisions a novel mix of recent AI techniques to combine the symbolic with the probabilistic, potentially enabling a closer modelling of human intuition. And cognitive scientist Gary Marcus, who in 2019 argued that neural networks would require the addition of symbolic computation to achieve common sense reasoning, recently co-founded Robust.AI in an effort to implement common sense in robots.

Perhaps the most notable AI advance heading into 2021 came out of the research and development of transformers, a new way of training machine learning systems without labelled data. The most prominent version of this, Google's GPT-3, could take a sample of writing and produce an entirely new text that made logical sense and retained the same style and tone. By internalizing the probabilities of paragraphs, the "generative pre-trained transformer" (GPT) could write new words as convincingly as if they were written by the original author. One can imagine an array of human-centered benefits, including ways to customize critical but complex messages, such as medical diagnoses and treatment plans, for a number of diverse audiences and people. And one could just as easily imagine more concerning uses: a company or government could intercept social media or other digital communications and massage the messages just enough to evoke a more "desirable" reaction, and do so without raising suspicion among its recipients.

What does all this innovation and upheaval mean? If nothing else, it means we have to get our arms around the opportunities and risks even faster than we'd previously argued. We've gone beyond the need for a digital Magna Carta. We now need to operationalize governance at the company and organizational level with data checks, algorithm audits, societal system impact testing, and enterprise governance that has teeth (i.e. with tangible rewards and penalties attached). Fortunately, we have some global forces working in our favor, and, because of the pandemic, we've never had a more opportune time to rebuild and reshape our societies and economies in more resilient, sustainable, and equitable ways. In the U.S., for instance, the Biden Administration's $2 trillion infrastructure plan will drive major federal investment into smart technologies and AI systems that improve lives and mitigate climate change. The increased scrutiny of the oligopoly power held by large digital platforms, particularly Facebook, Google, and Amazon, could help curb the now-common practices and systems that violate user's privacy and splinter communities. The European Union has advanced its regulatory responses to AI innovation even further, building on its General Data Protection Regulation (GDPR) with new draft rules for the European data market. Freshly released as we wrote this, the draft rules would govern how companies and governments can use artificial intelligence and the data that powers it. Even China has throttled back support for its huge homegrown digital platforms, particularly Tencent and Alibaba, the latter of which was fined $2.8 billion in April 2021 for violations of the country's anti-monopoly laws.

We can take heart in the fact that policymakers around the world have made moves to understand and check the largely unfettered growth of digital platforms, the power and valuation of which sky-rocketed during the pandemic. Collectively, regulatory efforts—if calibrated appropriately—could help address the societal divisions sown by these platforms, open up markets to increased competition, and curb the large players' ability to squelch innovation. Yet, the need to monitor and adjust these regulations will only grow more important as government hands grow heavier. Platforms continue to bring us new and often more accessible and affordable services than their analog counterparts, and we still need to encourage the entrepreneurs who will experiment and develop innovative AI-driven and data-fueled business models—the kinds of dynamism that can fuel significant economic and societal growth. Already back in 2017, PwC was forecasting that AI would add $15.7 trillion to the global economy by 2030, as companies improved productivity and consumers snapped up smarter products. Choking off that potential would be counterproductive.

Either way, the confluence of AI and advanced technology systems will exert an increasingly powerful influence on economic growth, international security and social stability. Global access to semiconductors, competition, and collaboration in advanced research fields such as quantum computing, and the embrace of AI-powered surveillance and other big-data applications have become central to today's geopolitical discourse. The National Security Commission on Artificial Intelligence (NSCAI) underscored that in its March 2021 report, urging U.S. lawmakers to advance innovation at home—through efforts to develop more U.S. talent, encourage domestic innovation, and protect American intellectual property—and to "build a favorable international technology order," as well. These efforts need to more clearly spell out areas of collaboration and competition between the U.S. and China, the two 800-pound gorillas of the data economy, because any meaningful progress on critical issues such as climate change, health, and the global data economy will be all but impossible without contributions from both nations. A more well-rounded strategy could define collective, global moonshots for economic growth and societal health—a worldwide Apollo program for the Cognitive Era.

One need only recall the international responses to the global Covid-19 pandemic to understand the importance of this mindset and the need to balance competition and collaboration. The same cross-border partnerships that led to the creation of mRNA vaccines could, put in the field of AI, lead to systems that help mitigate climate change, enhance our work, improve education, and transform international development. The competition between nations will undoubtedly remain, particularly between the U.S. and China, but not everything has to be competitive in Cold War terms. If policies, programs and products continually aim toward *symbio-intelligence*—using AI in ways that augment, not replace, human talents and skills—we can still avoid a generation of disenfranchised workers and expanding inequality. If we find ways to secure and diversify our data pools, we can still enhance our collective capacity to develop better, more-equitable, and increasingly beneficial AI systems.

These types of advancements serve every country's interest, and refusing to collaborate toward greater goals only risks a deeper erosion of trust—in our systems, in our institutions, in our allies around the world, and in our communities at home. After all, the urgency we feel today doesn't derive solely from the upheaval of the last three years, the rising market power of the digital barons, or the action or inaction of governments. It comes from a rapidly eroding sense of security, authenticity, and trust. Concerns about hacking and election security feed into concerns about authenticity. Deep fakes and other identity-masking capacities empowered by AI systems eat into our sense of trust. Every month, another app does an even better job of making someone look and sound like the president, the CEO, or your next-door neighbor.

The premium we put on trust, as the most valuable currency in life, has never been higher, and therein lies the true root of our urgency today. In 2018, we firmly believed that we needed to build guardrails for the cognitive economy and society, so we could progress at entrepreneurial speed, capitalize on its benefits, and mitigate its risks. We could see the traffic gathering on the road ahead, and we urged business leaders, researchers, and policy makers to get out ahead of it. Now, the cognitive society is here, evolving at ever faster speeds, and we still don't have the guardrails we need to create and recover the much-needed trust in the road ahead. Let's build forward better!

—Olaf Groth and Mark Nitzberg
Berkeley, California

Index

INDEX

282